GUIDE TO WALLPAPER AND PAINT

McGRAW-HILL BOOK COMPANY

New York St. Louis San Francisco Auckland Bogotá Düsseldorf
Johannesburg London Madrid Mexico Montreal New Delhi Panama
Paris São Paulo Singapore Sydney Tokyo Toronto

McGraw-Hill Paperbacks Home Improvement Series

Guide to Plumbing

Guide to Electrical Installation and Repair

Guide to Roof and Gutter Installation and Repair

Guide to Wallpaper and Paint

Guide to Paneling and Wallboard

1 2 3 4 5 6 7 8 9 0 SMSM 8 3 2 1 0

Library of Congress Cataloging in Publication Data
Main entry under title.

Guide to wallpaper and paint.

(McGraw-Hill paperbacks home improvement series)
Originally issued in 1975 by the Automotive-Hardware Trades Division of the Minnesota Mining and Manufacturing Company under title: The home pro wallpaper and paint guide.
1. Paper-hanging — Amateurs' manuals. 2. House painting — Amateurs' manuals. I. Minnesota Mining and Manufacturing Company. Automotive-Hardware Trades Division. The home pro wallpaper and paint guide.
TH8441.G84 1980 698.6 79-14723
ISBN 0-07-045961-4

Front cover photo: Blietz Valenti, Inc.

Back cover photo, top, courtesy of Stauffer Chemical Company.

Design: Margot Gunther. Wallcoverings: Stauffer Miniatures collection. Furniture: Plymwood, a division of Sugar Hill. Lamp: Stiffel. Needlepoint: Columbia Minerva. Handmade rug: Pilgrim's Progress, a New York retail store.

Back cover photo, bottom, courtesy of Pittsburgh Paints.

Contents

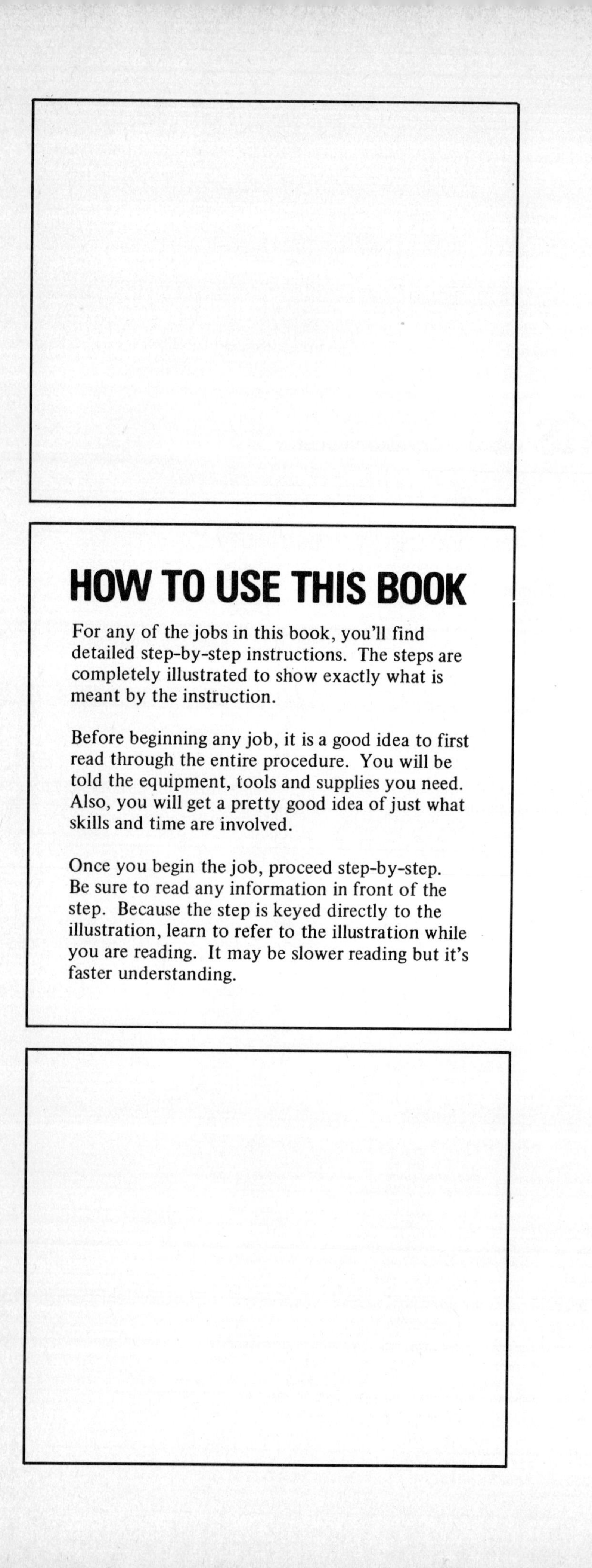

HOW TO USE THIS BOOK

For any of the jobs in this book, you'll find detailed step-by-step instructions. The steps are completely illustrated to show exactly what is meant by the instruction.

Before beginning any job, it is a good idea to first read through the entire procedure. You will be told the equipment, tools and supplies you need. Also, you will get a pretty good idea of just what skills and time are involved.

Once you begin the job, proceed step-by-step. Be sure to read any information in front of the step. Because the step is keyed directly to the illustration, learn to refer to the illustration while you are reading. It may be slower reading but it's faster understanding.

WALLPAPERING

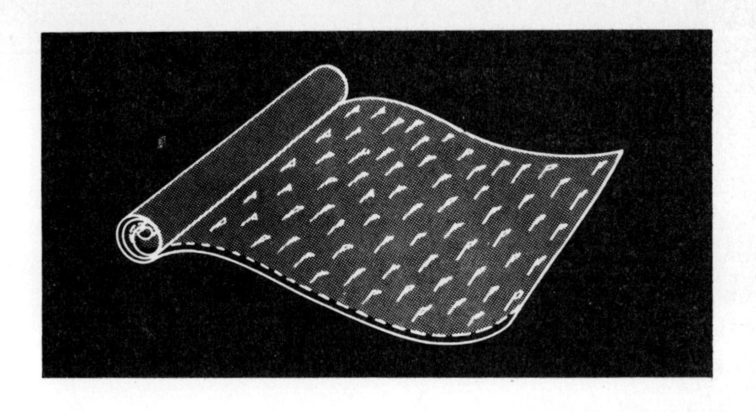

The term wallpapering applies to hanging any type of wallcovering – paper or not. It includes vinyls, foils, grass cloth, cork and any of the other materials which are available at your wallpaper dealer.

Wallpapering can be done with professional results by almost anyone willing to plan carefully and work slowly. It is not a difficult task.

The selection of wallpaper color and pattern is pretty much a personal thing. Books about interior decorating are full of ideas regarding the use of wallpaper. In addition, many wallpaper dealers can give you valuable advice about choosing wallpaper to achieve different effects – brightening a room or wall, lengthening a room, coordinating carpet and furniture colors and styles.

The selection of wallpaper material is a matter of function. Vinyl wallcoverings should certainly be used in kitchens, bathrooms or hallways. It is extremely durable and moisture and grease resistant. In an entry way, durability is not a major consideration but dramatic appearance may be. Foils, flocks or grass cloth may be the answer to a decorating problem here.

This book shows you how to hang any wallcovering that you choose. Before starting with any of the procedures for hanging wallpaper, be sure to read the following paragraphs.

Wallpaper Colors

The colors in a wallpaper will vary slightly between different runs. If you have to reorder wallpaper, there is a good chance that the new order will not color match the previous order. Therefore, estimate your needs carefully and order enough wallpaper at one time to complete your entire job.

Wallpaper Patterns

Blank stock is an unpatterned wallpaper commonly applied to a wall to provide a smooth surface for foil wallpaper and other special applications. All other wallpaper is patterned.

Some patterns, many of the vertical stripes for example, require little or no matching. The paper can be cut almost to the exact length required. Since only an inch or two extra length for trimming is required of each strip very little waste results. The same is true of randomly patterned papers such as grass cloth or burlap.

Other patterns can result in considerable waste. A clue to the amount of waste involved can be gained from the size of the repeat. This information is provided in the wallpaper book from which you are making your selection.

The book will give the interval with which each pattern is repeated. For example, the book may state that a certain pattern is repeated every 19 inches. Assume that the distance between the baseboard and the ceiling of the room you want to paper is 94 inches. The 19 inch pattern will go into the 94 inch wall 5 times – with 1 inch left over for trimming. This pattern would result in very little waste.

You may find that another pattern is repeated every 18 inches. For the same wall, this pattern would repeat 6 times ($94 \div 18$) with 14 inches left over. The 18 inch pattern results in 14 inches of trim for each strip required. Unless you can find a place (over a window or door, for example) to use these remnants, you will have a lot of waste.

Common Cautions You Should Know

1. Do not paper directly on new wallboard or plaster. Apply flat oil-based paint, wall size or oil-based primer.

2. Do not paper over wallpaper which is in poor condition. Repair damage, remove loose paper or remove all old paper.

3. If you use lightweight or porous paper, observe the following cautions:

   - Do not paper over dark paper or patterns. They may show through the new paper. Lighten background with paint or blank stock.

   - Do not paper over wallpaper if the ink comes off. It could bleed through the new paper. Seal the old paper with oil-based primer or wall size.

   - Do not paper over dark painted walls. They could show through. Lighten them with paint.

4. If you are also painting, complete all painting before beginning papering.

5. Read special hints for hanging your particular type of wallpaper on the following pages:

Type of Wallpaper	Page
Vinyls	3
Foils	3
Grass Cloth, Hemp, Burlap, Cork	4
Flocks	4
Pre-pasted Wallpaper	5

▶ Vinyls

Three kinds of vinyl wallpaper are commonly available:

- Vinyl laminated to paper
- Vinyl laminated to cloth
- Vinyl impregnated cloth on paper backing

These vinyl wallpapers are the most durable wallpapers made. They are easy to clean (scrubbable) and are very resistant to damage.

Most vinyl wallpapers are nonporous.

CAUTION

Some wallpapers are vinyl-coated only. They are not particularly wear-resistant, grease resistant or washable. Do not confuse them with vinyl wallpapers.

Preparation for Hanging

1. Any wallpaper on surface to be papered should be removed. It is probably glued to the wall with wheat paste. Because vinyls are nonporous, moisture from adhesive will be sealed in and cause wheat paste to mildew. Also, sealed in moisture can soak through old wallpaper and cause it to peel loose from wall.
2. It is recommended that sizing be applied to surfaces before papering. Sizing provides a good bonding surface. It also makes it easier to slide wallpaper strips into alignment for matching and making seams. The best size is a coat of the same vinyl adhesive you use for applying the vinyl wallpaper. Be sure that any size that you use is mildew resistant.

Adhesives

Use mildew resistant adhesives only. It is recommended that vinyl adhesives be used.

Paperbacked vinyl sometimes tends to curl back from the wall along edges at seams. Seams can then be difficult to finish. Use of vinyl adhesive prevents this problem.

Hanging

1. Vinyl wallpaper stretches if pulled. If it is stretched while it is applied, hairline cracks will appear at seams when the wallpaper shrinks as it dries. Therefore, be careful to avoid stretching vinyl wallpaper.
2. Some persons find that a squeegee works better than a smoothing brush for smoothing vinyl wallpaper.

▶ Foils

There are two types of foils:

- Simulated metallic
- Aluminum laminated to paper

All foils must be handled carefully. Be careful not to fold or wrinkle foils because creases cannot be removed.

Because of their reflective surface, foils will magnify any imperfections on the surface to which they are applied.

Foils are nonporous. They do not "breathe".

Preparation for Hanging

1. Any wallpaper on surface to be papered should be removed. It is probably glued to the wall with wheat paste. The wheat paste is likely to mildew when covered by foil. Also, the moisture sealed in by foil could soak through old wallpaper and cause it to peel loose from wall.
2. Do not apply a cereal-based sizing to the wall. It could mildew also.
3. Sand wall lightly with fine sandpaper to remove texture and imperfections. Any rough spots will be magnified when covered by foil.
4. It is generally recommended that the surface first be covered with blank stock. Blank stock is an unpatterned wallpaper available from your wallpaper dealer. Blank stock serves two purposes – it will help smooth the surface and it will absorb moisture from the adhesive used for the foil, thus speeding up drying time. Apply blank stock with the same adhesive used for applying the foil wallpaper.

Adhesives

Use mildew resistant adhesives only. It is recommended that vinyl adhesives be used. Because foils are nonporous, the adhesive is sealed from air and is very slow to dry. Vinyl adhesives are relatively fast drying.

Wheat paste could mildew. Never use wheat paste with foil.

Hanging

1. Be careful when handling foil. Creases cannot be removed.
2. Some foils must be hung dry. Paste is applied to the wall and the foil is positioned on the paste. Special instructions like these would be enclosed with your wallpaper.

▶ Grass Cloth, Hemp, Burlap, Cork

These materials are generally mounted on a paper backing.

The patterns and textures are generally random and require no matching. In fact, for grass cloth even the color will change slightly between one part of a roll and another part.

These materials are not washable.

Preparation for Hanging

1. The paper backing on which these materials are mounted can be weakened from oversoaking with paste. Therefore, it is recommended that the wall first be covered with blank stock to help absorb moisture from the paste. Blank stock is an unpatterned wallpaper available from a wallpaper dealer. Get his advice regarding the use of blank stock for your situation.

2. If you decide not to use blank stock, you should apply sizing to the wall. Sizing provides a sealed surface which makes it easier to slide the wallpaper into place when making seams and corners.

3. Color and texture of grass cloth varies between rolls. It is a good idea to first cut all full-length strips and arrange them for best appearance. Then stack them in this order for pasting.

4. Because patterns do not require matching, they result in little waste. When cutting strips, allow only 1 or 2 inches extra length for trimming.

5. Sometimes the edges of grass cloth may be ragged. If this happens, you may wish to trim 1/2 inch off the edges before hanging. This trimming will give you a good, sharp edge for seams. For best results, trim the edges after the paper is pasted.

Adhesives

Wheat paste or liquid cellulose.

Hanging

1. Be careful not to oversoak backing with paste. Paper backing can be weakened and allow the surface to separate. Therefore, rather than paste several strips ahead, paste a strip and hang it before pasting the next strip.

2. These materials may be difficult to cut when wet. The job of trimming will be easier if you align the top end of the strip to the ceiling and do all your trimming at the bottom end of the strip. Mark it at the baseboard and cut with a scissors.

3. To align seams, push wallpaper gently with palms of hands. Do not use a smoothing brush for aligning. Paper can be damaged by rubbing. This paper can be smoothed to the wall without damage with a paint roller.

4. Do not use a seam roller on seams. Press seams into place with a soft paint roller. Also press into place using fingers and damp cloth.

5. Don't rub surface of wallpaper. Remove excess paste by wiping gently or blotting with a damp sponge.

▶ Flocks

Flock is made of nylon or rayon.

Flock is available on paper, vinyl or foil wallpapers.

Flock is washable – but can be damaged by rubbing.

Preparation for Hanging

Flocks cause no special preparation problems. Read preparation information which applies to your wallpaper material – vinyl, foil.

Adhesives

Use adhesive recommended for your wallpaper.

Hanging

1. Flocks can be damaged and flattened by rubbing and pressure. Therefore, try using a paint roller or squeegee (available at wallpaper dealer) rather than a smoothing brush to smooth wallpaper.

2. Do not use a seam roller on seams, edges or ends. Instead, pat down edges with a damp sponge or cloth. Some persons use a new soft paint roller for this purpose.

3. After hanging a strip, wash it in downward direction with a damp sponge. Then, with a clean damp sponge, fluff up all flock with upward strokes to lay all nap in same direction. Some small fibers will come loose during washing. This is no cause for concern.

▶ Pre-pasted Wallpaper

This wallpaper has water-soluble paste applied at factory.

After soaking strip in water according to manufacturer's instructions, strip is ready for hanging.

Pre-pasted wallpaper is available in vinyl and paper materials.

Preparation for Hanging

Pre-pasted wallpapers cause no special preparation problems. Read preparation information which applies to your wallpaper material.

Adhesives

Follow manufacturer's instructions for wetting adhesive.

Hanging

1. When soaking strip in water tray, weight strip with butter knife or other dull, heavy object to make it stay under water and to facilitate unrolling strip.

2. There are two different ways to soak and hang strips:

   a. Place water tray alongside wall at position strip is to be hung. Grasp top end of strip and lift into position on wall. Note – if strip is made long enough so that bottom end of strip remains in tray after top end is in position on wall, excess water will run off into tray rather than onto floor.

   b. Place tray next to work table. Lift strip onto table, fold strip and let paste cure. Unfold strip and hang.

ESTIMATING WALLPAPER, TRIM AND PASTE

▶ How Much Wallpaper

This section provides instructions for estimating the amount of wallpaper needed to cover walls and ceilings in rooms of different shapes and sizes.

Always plan on having some extra wallpaper; it may be needed for several reasons:

- A strip of wallpaper may be ruined in handling.
- Light switch covers, electrical outlet covers, and other surfaces in a room may be included in the job, requiring additional wallpaper.
- Future repairs may become necessary for which it is important to have wallpaper of the original batch. The same wallpaper, if bought later, may not exactly match that on your wall for color.

The amount of wallpaper required to cover a surface depends not only upon the size of the surface but the wallpaper pattern itself.

To make your own estimate, you should know the following facts about wallpaper:

Wallpaper is sold in a variety of widths. American-manufactured wallpaper rolls are 36 square feet, regardless of the width.

The length of each roll depends on the width. The wider the roll, the shorter the length, and therefore fewer strips are available in the roll for hanging. European wallpaper varies in square footage. However, it usually lists an American roll equivalent on the package.

When estimating the amount of wallpaper needed, count on 30 square feet of coverage per roll, allowing 6 square feet for waste. The exact amount of waste depends upon the pattern you select.

When estimating wallpaper for surfaces that are interrupted by doors, windows, and fireplaces, make your estimate for the entire surface. Then reduce your estimate by 1/2 roll for each door, window, and fireplace, or 15 to 20 square feet each.

If you estimate your wallpaper in this way, you should have enough wallpaper to complete the job plus some left over.

The surest method of getting the correct amount of wallpaper is to have your wallpaper dealer estimate it. He is an expert. After you have selected a wallpaper pattern, he can tell you exactly how much you need if you bring him the following sketches and measurements.

ESTIMATING WALLPAPER, TRIM AND PASTE

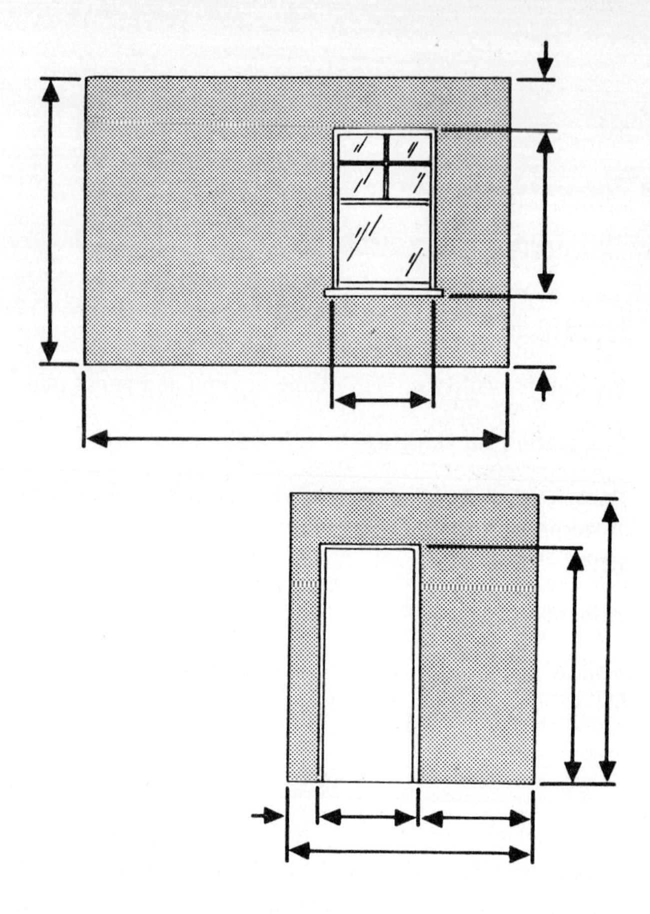

How Much Wallpaper

A separate sketch should be drawn for each surface to be papered.

1. Draw sketch of each surface to be papered.

2. Add to the sketch all windows, doors, fireplaces, etc., which are not to be papered.

Most ceiling areas can be measured at floor level. If a ceiling is sloped or has a complicated shape, it may be necessary to measure it directly.

3. Measure and record all dimensions on sketch.

▶ How Much Paste

The amount of paste needed for a wallpapering job depends on:

- Number of wallpaper rolls
- Type of wallpaper

When wheat paste is used, one pound of dry mix is generally enough to hang six to eight rolls of wallpaper. More dry mix than recommended by the label instructions may be needed. Following label instructions sometimes results in a paste which is too thin.

Some dry mix pastes contain mildew resistant additives for use with coated or vinyl wallpaper. Always use mildew resistant pastes or glues with coated wallpaper.

When vinyl adhesive is used, the weight and type of backing on the vinyl determines the amount of adhesive needed. One gallon of vinyl adhesive is generally enough to hang two to four rolls of wallpaper.

Ask a wallpaper dealer to recommend the amount and type of paste or adhesive needed for your specific wallpapering job.

▶ How Much Trim

Trim can be used as decorative border, or to lessen the effects of any sudden change in patterns and colors between ceilings and walls.

Trim is sold by the yard. Therefore, when measuring for trim, round measurement to the next higher yard.

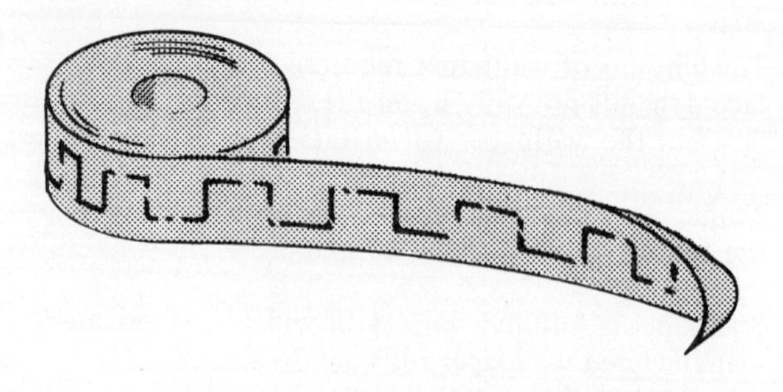

▶ Preparing Wall and Ceiling Surfaces

New wallpaper is sometimes hung over old wallpaper. Whether the job is successful and gives a good appearance depends upon the number of layers of old wallpaper and the condition of the old wallpaper.

It is generally recommended that all old wallpaper be removed before hanging new wallpaper. Although this used to be a very difficult, time-consuming job, special equipment and chemicals are now available which make the job much easier.

The procedures for preparing the walls and ceilings are as follows:

1. If surfaces are papered, remove old wallpaper. Go to Page 44 to remove wallpaper.

2. Inspect surfaces for cracks and holes.

If any cracks and holes are found, they must be repaired and sanded.

3. Check condition of painted surfaces.

If there is dirt, grease or wax, it must be removed. Following manufacturer's instructions, mix a solution of trisodium phosphate (T.S.P.). Wash surfaces with solution. After cleaning painted surface, go to next section (below) to apply sizing.

If paint is shiny and glossy, wallpaper may not stick well to the surface. This condition can be corrected by scrubbing with a solution of T.S.P. or by applying a commercial deglosser. Consult a paint dealer for materials and recommendations. After removing shine or gloss, go to next section (below) to apply sizing.

4. If surface is not painted, it must be sealed before applying sizing.

Apply an oil-based sealer or a primer to unpainted surfaces such as new wallboard, new plywood or new plaster. After sealing surfaces with sealer or primer, go to next section (below) to apply sizing.

▶ Applying Sizing

It is generally recommended that sizing be applied to surfaces before papering them. Sizing acts to seal the surface, preventing it from absorbing water from the wallpaper paste or glue. It allows the paste or glue to dry properly. Sizing also provides a roughened surface to which wallpaper paste or glue can stick firmly.

Wallpaper sizing is inexpensive and easy to apply.

Two types of sizing are available:

- Wallpaper paste, mixed according to manufacturer's instructions. It will act as a sizing agent when applied to a surface and allowed to dry.

- Wallpaper sizing mixed according to manufacturer's instructions. It is especially made for wallpaper.

Using paintbrush or roller, apply sizing to surface. Allow sizing to dry completely.

PREPARATION FOR HANGING WALLPAPER

▶ Preparing Work Area

Most amateurs will find that hanging wallpaper is a slow job at best in order to get professional looking results. A few minutes spent organizing the work at the beginning will save time, mess and tempers.

An empty room makes the best work area. However, it is seldom practical to do this. Move what furniture you can out of the room. Move the rest away from your work area so that you have as much room as possible for using your ladder and handling strips of pasted wallpaper. Nothing is so frustrating as cramped space.

Furniture that can't be removed should be protected. Spread drop cloths or paper on floors and carpets.

Take down all removable hardware that will be in your way. Such hardware includes:

- Cover plates from electrical switches and outlets. Go to Page 36 for instructions.

WARNING

Removal of cover plates will make it easier to touch "hot" wires. Therefore, if children are around, remove cover plates only at the time necessary for papering around the switch or outlet.

- Light fixtures. Go to Page 37 for instructions.

Since electrical power will be off, you may find it convenient to remove fixtures only at the time necessary for papering around the fixture.

- Curtain rods and brackets, drapery rods and brackets, picture hooks.

Screw holes will be difficult to relocate after they are covered with wallpaper. To save time in relocating them and to avoid having to drill new holes do the following:

1. Place a toothpick in each hole.
2. Remove toothpick when papering over it.
3. Push toothpick through paper into hole before going on to next strip.

Preparing Work Area

Bring all of the tools and supplies you will need into your work area. Read Pages 9 and 10 to make sure you have everything. Arrange a place to cut and paste wallpaper as follows:

1. Set up pasting table [1] or area away from surfaces to be wallpapered.

CAUTION

Do not use newspaper to cover pasting table [1]. Newspaper ink may soil wallpaper.

2. Cover pasting table [1] with clean, plain paper [2] such as freezer wrap or brown wrapping paper. Tape paper in place.

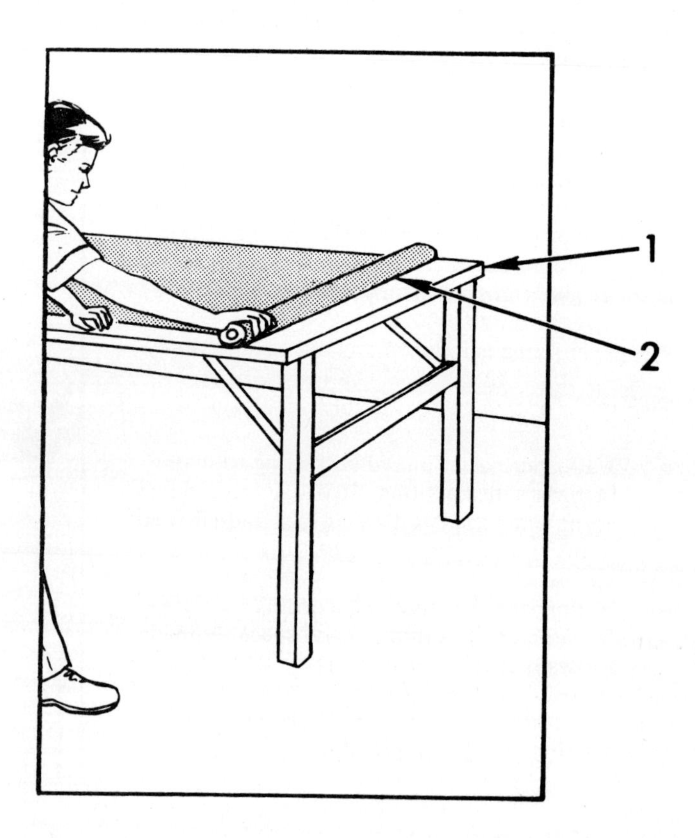

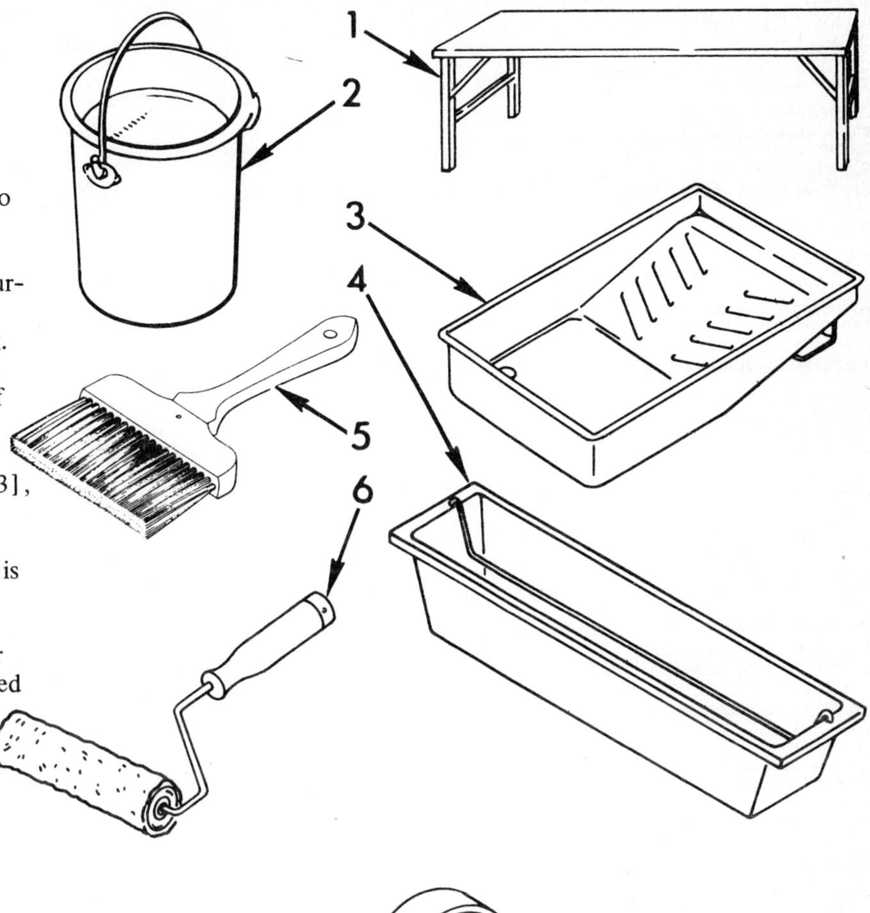

▶ **Tools and Supplies**

The following tools and supplies are needed to hang wallpaper:

- A pasting table [1] or other flat, hard surface for cutting and pasting wallpaper. Pasting table must be at least 6 feet long. It can either be rented, or it can be constructed by placing a half-width sheet of plywood on two or three sawhorses
- Two buckets [2], pails, or roller trays [3], one for paste, one for clear water
- A water tray [4] if prepasted wallpaper is used
- A pasting brush [5], calcimine brush, or paint roller [6] to apply paste if unpasted wallpaper is used

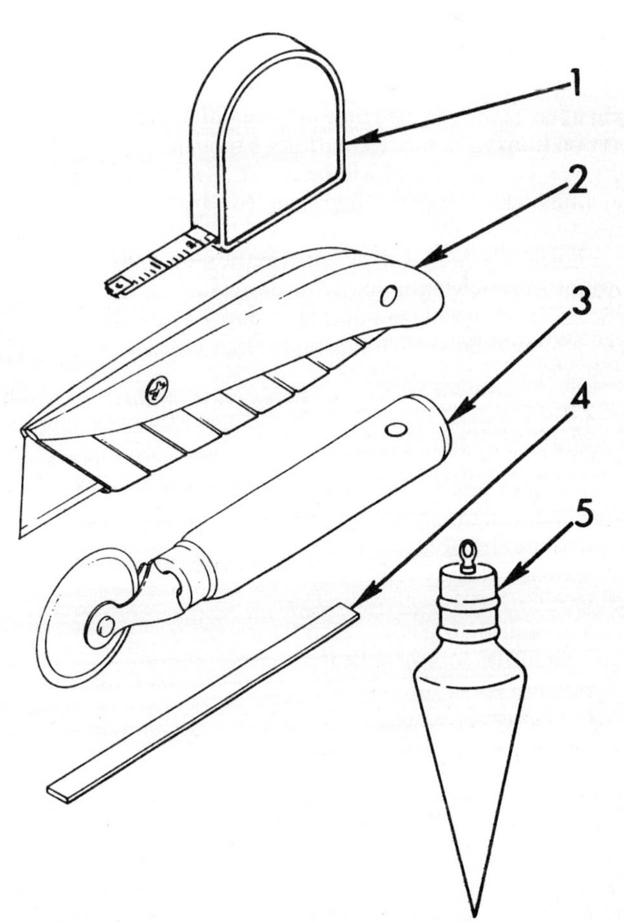

Tools and Supplies

- A flexible metal rule [1] to measure wallpaper lengths and widths
- A trimming knife [2] to cut wallpaper. This can either be a sharp knife or a razor blade and holder.
- A trimming wheel [3] to cut wallpaper. Two kinds, the perforated wheel and the knife-edge wheel, are available.
- A straightedge [4] to aid in cutting straight lines. A metal edge is recommended but any long, straight object can be used.
- A plumb [5] to give a true, straight, vertical line. A plumb can be constructed of a piece of string, some colored chalk and any object suitable for a weight.

HANGING WALLPAPER

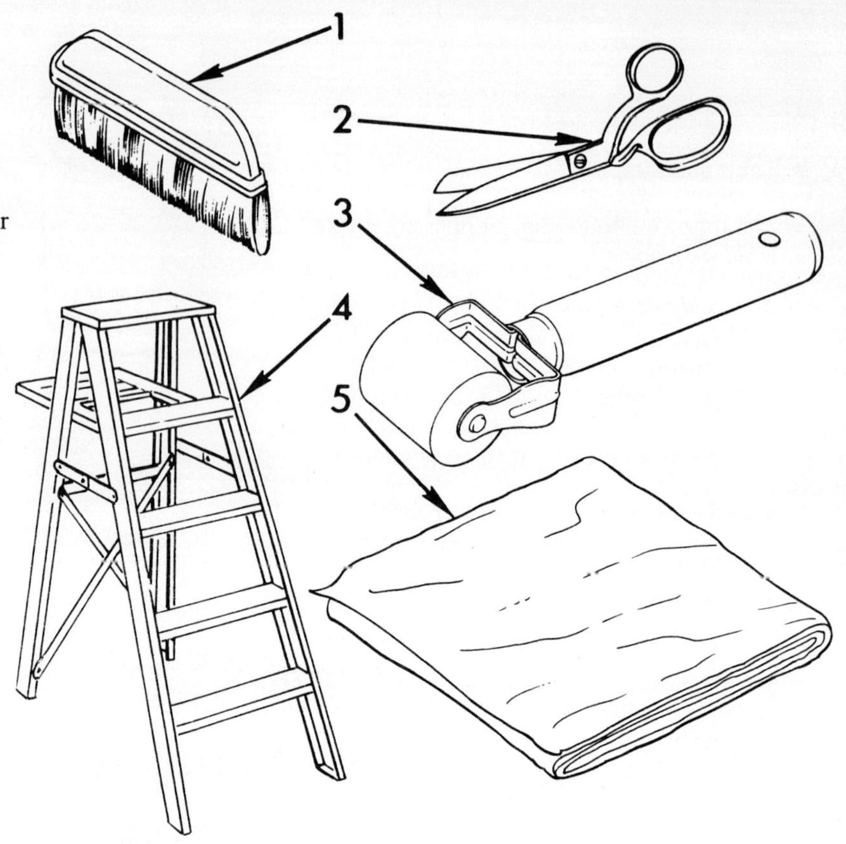

Tools and Supplies

- A smoothing brush [1] to smooth wallpaper against wall and to remove air bubbles and wrinkles. Brush should be 12 inches wide, with bristles that are firm but soft enough not to scratch wallpaper.
- Shears [2] or heavy-duty scissors to cut wallpaper
- A seam roller [3] to make tight joints at seams and edges
- A step ladder [4] or stool to reach high places. If ceiling is to be wallpapered, two ladders and scaffold planks should be used.
- Drop cloths [5] to protect floors and furniture not removed from room

▶ Selecting Position for First Strip

Wallpaper, paint and paneling are used in many combinations in home decoration. Frequently, only one wall of a room will be papered while the remaining walls are painted or paneled. In other cases, two, three or all four walls will be papered. The choice of which wall or combination of walls to paper is somewhat a matter of individual preferences. However, if you decide to paper one wall only, it should be:

- The wall opposite the most frequently used entrance to the room, or
- The wall that you most commonly view when using the room. For example, this may be the fireplace wall or picture window wall in a living room, or
- The wall that needs special treatment to lighten the room, make it appear larger or provide visual interest. An example would be the use of scenic mural on a wall to make the room appear larger.

Just as there are rules for selecting which wall to paper, there are rules for selecting where to start papering the wall.

If all four walls are to be papered, there are not only rules for where to start papering, but also where to finish the papering.

If all four walls are to be papered there will almost always be one place where the paper cannot be matched. Therefore, you should plan to locate this mismatch in the least noticeable place in the room. The best place for the mismatch is the least noticeable corner. However, sometimes a door opening or built-in cupboards or bookshelves can make a break in the pattern so that a mismatch will not be noticeable.

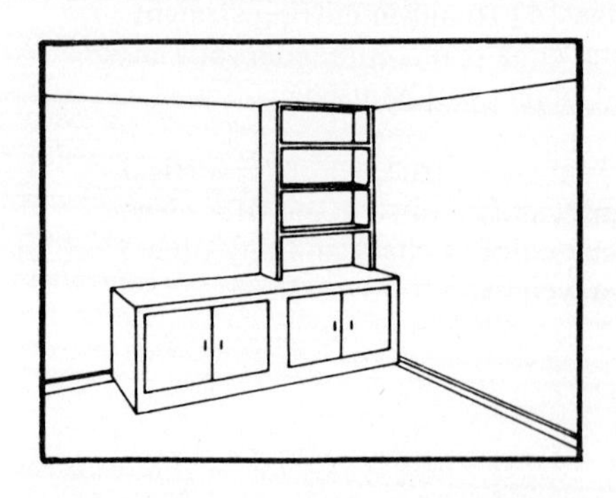

Selecting Position for First Strip

As a general rule, you should begin your papering on the wall which is most noticeable and end your papering in the corner which is least noticeable. After placing the first strip of wallpaper on the wall, work from the right or left to the least noticeable corner. Then work from the opposite direction to the least noticeable corner.

The following instructions show how to locate the first strip for several kinds of situations:

- Wall with two windows. Page 22.
- Wall with fireplace or picture window. Page 13.
- Wall with no unusual features (Start papering at a corner, door frame or window frame). Page 13.
- Ceilings (If the ceiling is to be papered, always paper the ceiling before papering the walls). Page 14.

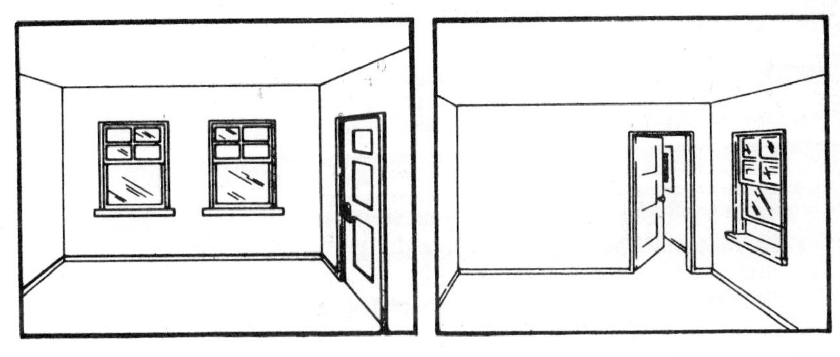

▶ Making a Plumb Line

A plumb line is a linc that is vertical. If the first strip of wallpaper which is hung on a wall is not vertical, the pattern along the top of the wall will be slanted [1]. The room will appear slanted. Therefore, it is necessary that a plumb line be marked on each wall. The first strip of wallpaper on a wall must be aligned to the plumb line.

A plumb line is usually made from ceiling to floor but can be made at any vertical surface. When making a plumb line, the string [2] should be long enough to allow the plumb [4] to swing freely within 2 inches of floor.

1. Where plumb line mark is desired, place a tack [1] in wall 1-1/2 inches below ceiling. Tie string [2] to tack. Tie plumb [4] or other weight to string.
2. When plumb stops swinging, place mark [3] on wall 2 inches above baseboard [5]. Remove plumb.
3. Rub string [2] with colored chalk.
4. While holding string [2] tight between tack and mark [3], pull string straight back from wall. Release string.
5. Chalk should mark straight, vertical line on wall. Remove tack and string [2].

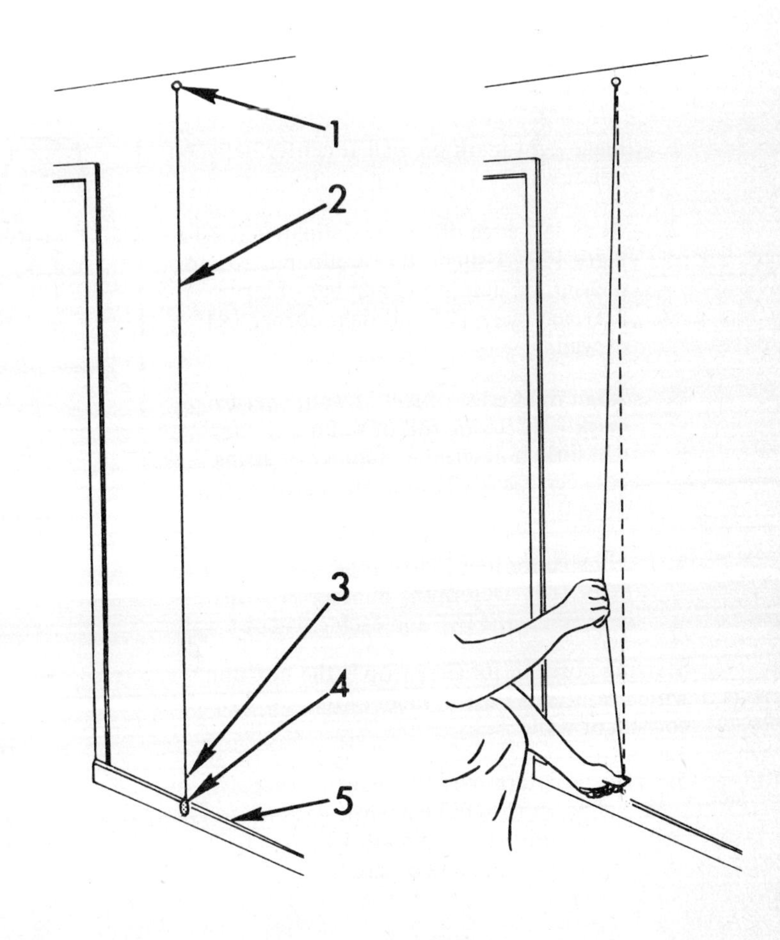

▶ Marking Position for First Strip – Wall with Two Windows

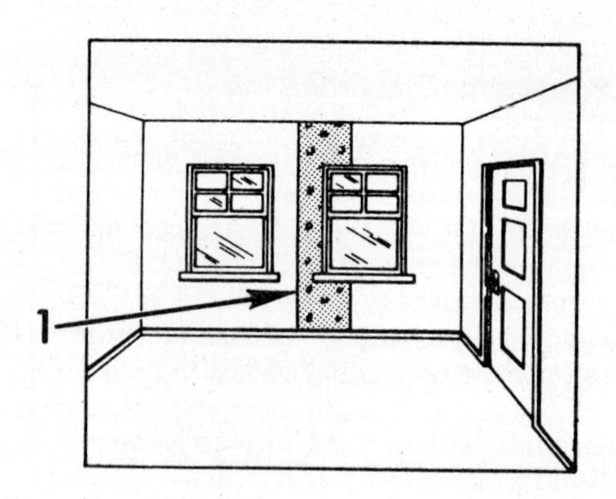

If the wall that you select to begin papering the room has two windows, locate the first strip between the two windows. The windows will then be framed similarly by the patterns on the wallpaper.

You should determine whether to position the first strip:

- So that the edge [1] of the strip is positioned at the center between the windows,
  or
- So that the centerline [2] of the strip is positioned at the center between the windows.

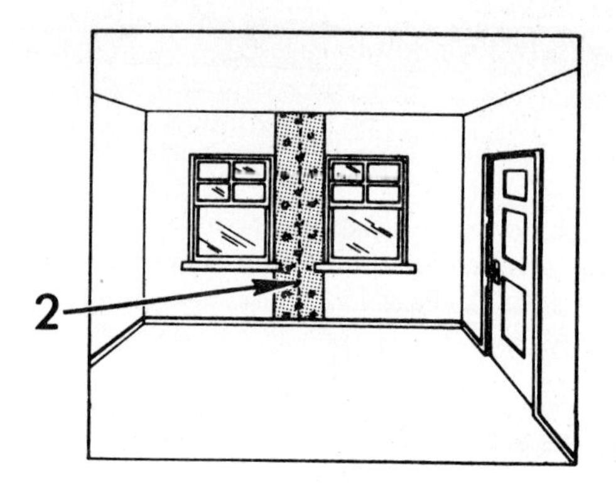

To do this, determine which location best avoids your having to work with very narrow strips of wallpaper at the corners of the wall. It could be difficult to work with strips which are narrow (6 inches or less wide) with some types of wallpaper materials. To find the best location for starting the first strip, go to Step 1.

Marking Position for First Strip – Wall with Two Windows

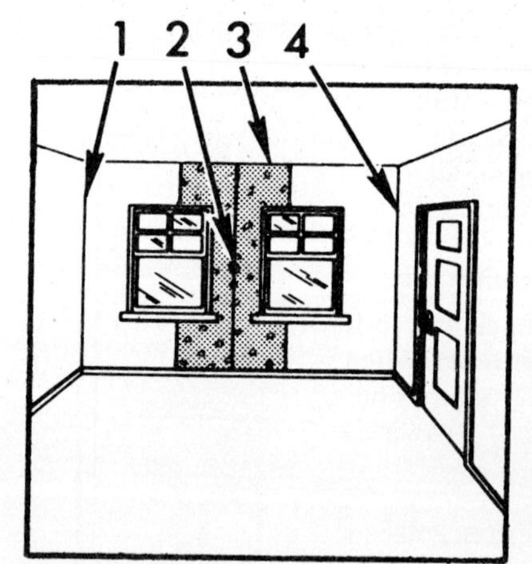

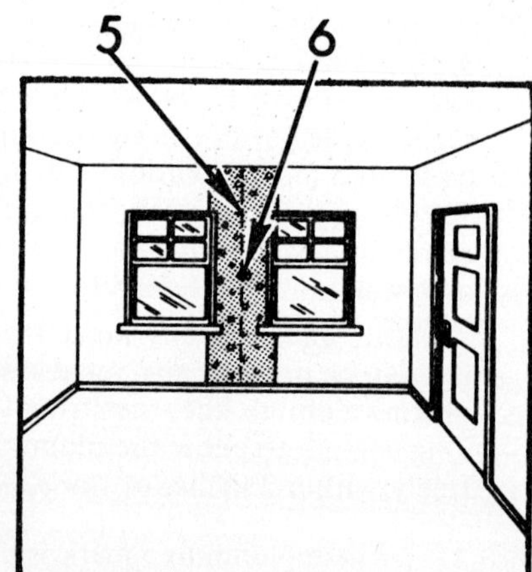

1. Place light mark on wall at center between windows.
2. Align left edge of roll of wallpaper [3] to mark [2]. Using roll of wallpaper [3] for measuring, determine number of strips between mark [2] and right corner [4] of wall.
3. Align right edge of roll of wallpaper to mark [2]. Using roll of wallpaper for measuring, determine number of strips between mark [2] and left corner [1] of wall.
4. Align centerline [5] of roll of wallpaper to mark [6]. Determine number of strips between mark [6] and each corner of wall.

Starting position for first strip is the position which results in least number of narrow strips at corners of wall.

5. If it is determined that edge of first strip of wallpaper should be aligned with mark [2], make plumb line on mark [2]. Go to Page 11 for making plumb line.
6. If it is determined that centerline of first strip of wallpaper should be aligned with mark [6], make plumb line 1/2 width of wallpaper away from mark [6]. Go to Page 11 for making plumb line.

▶ Marking Position for First Strip — Wall with Fireplace or Picture Window

On a wall with a fireplace or picture window, the first length of wallpaper should be located above the fireplace or window.

The strip can be located so that

- One edge [1] of the strip is positioned at the center of the fireplace or picture window, or
- The centerline [2] of the strip is positioned at the center of the fireplace or window.

The choice of starting place can result in easier papering of the corners or a better appearance of the wallpaper pattern at the corners. Place mark on wall at centerline of fireplace or window. Then go to Page 12 (bottom) for procedures for determining best alignment position for first strip.

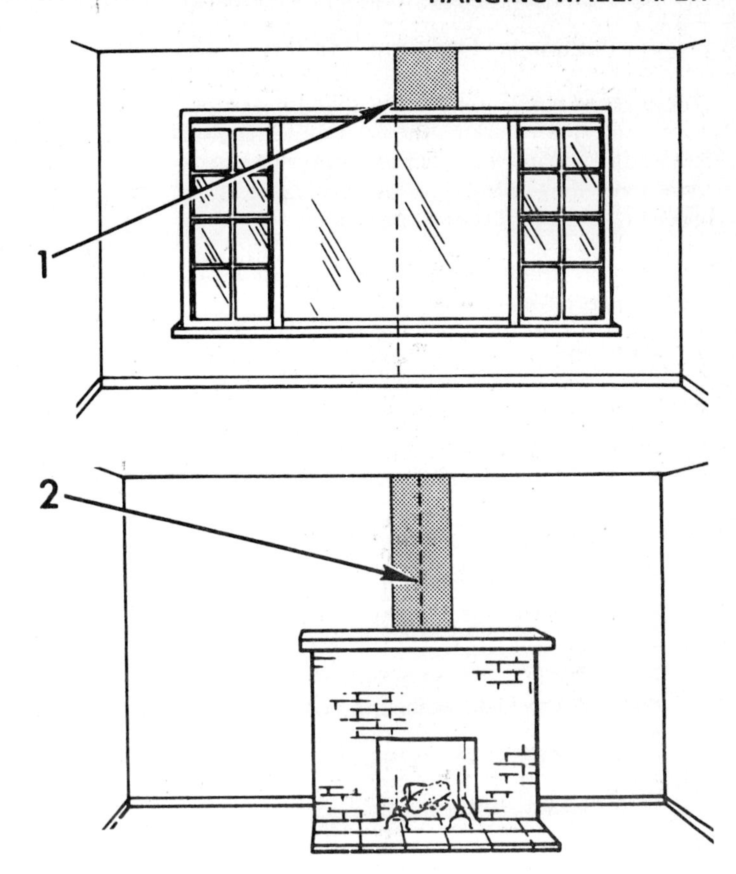

▶ Marking Position for First Strip — Corner, Door Frame or Window Frame

If you begin at a place other than a fireplace, picture window or wall with two windows, a corner [1], door frame [2] or window frame [3] is usually the best place to begin.

1. Measure and record width of wallpaper. Subtract 1/2-inch from width.

   Example: 18 inches minus 1/2-inch = 17-1/2 inches

2. Beginning at corner [1], door frame [2] or window frame [3], measure distance figured in Step 1. Place mark on wall.

3. Make a plumb line at mark. Go to Page 11 for making a plumb line.

If you are papering around a corner, you must make sure that wallpaper strip will overlap corner by no less than 1/2-inch. Because corners are usually not straight, you should check distance between plumb line and corner at three places to ensure that overlap is not less than 1/2-inch.

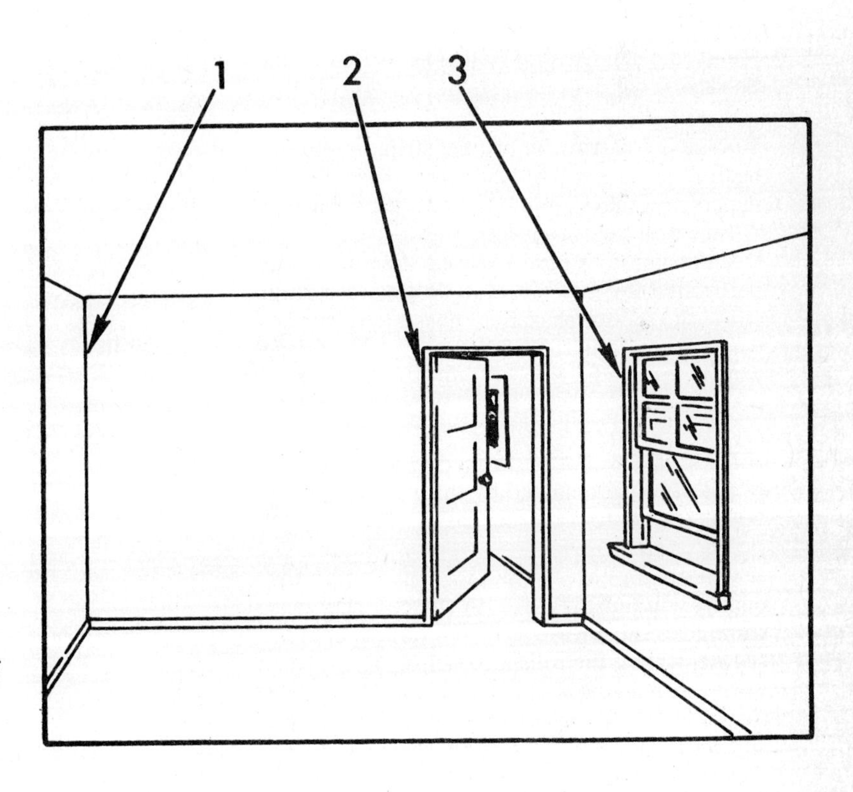

HANGING WALLPAPER

▶ Marking Position for First Strip – Ceiling

It is easier to handle short strips than long strips. Therefore, when marking the position for the first length of wallpaper on a ceiling, always be sure to mark the ceiling in the shortest direction. Cut and hang strips in this short direction.

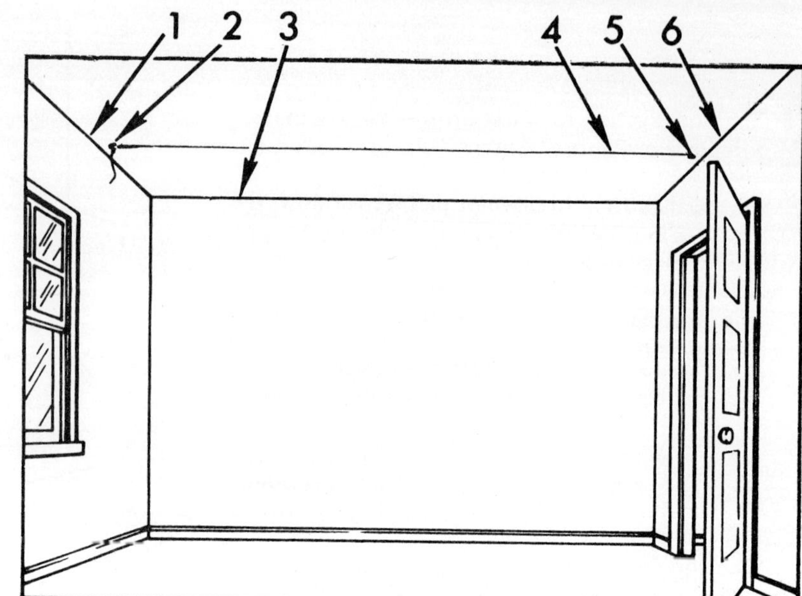

1. Measure and record width of wallpaper. Subtract 1-inch from width.

   Example: Width of wallpaper is 18 inches. 18 inches minus 1 inch = 17 inches.

2. Beginning at wall [3], measure distance figured in Step 1. Place mark [2] on ceiling 1-1/2 inches from wall [1].

3. Beginning at wall [3], measure distance figured in Step 1. Place mark [5] on ceiling 1-1/2 inches from wall [6].

4. Place tack in ceiling at mark [2]. Tie string [4] to tack. Rub string with colored chalk.

5. While holding string [4] tight between tack and mark [2], pull string straight back from ceiling. Release string. Chalk should mark straight line on ceiling.

6. Remove tack and string.

▶ Measuring and Cutting Wallpaper – Removing Selvage

Selvage is a narrow, unprinted strip on one or both edges of a roll of wallpaper. Very few wallpapers nowadays are purchased with selvage.

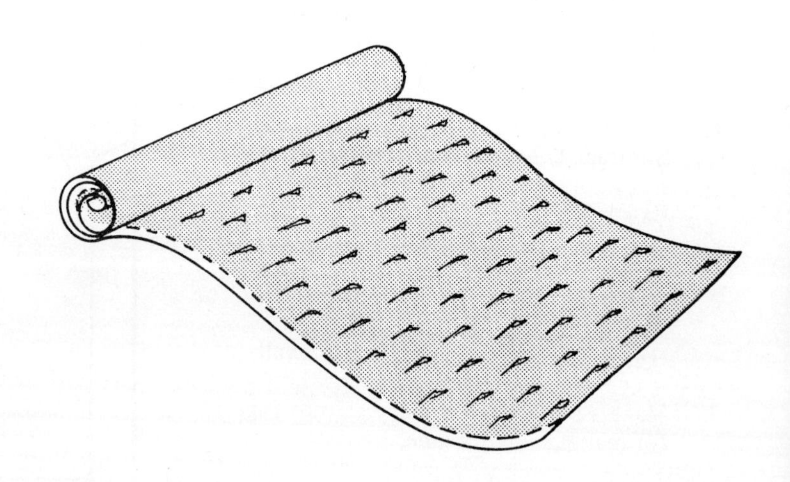

If wallpaper does have a selvage strip, it is often perforated for easy removal. In this case simply strike the selvage sharply against a hard surface while turning the roll of wallpaper. The selvage will eventually be torn off without damaging the pattern.

If the selvage is not perforated, it must be cut off. It can often be removed by the wallpaper dealer with a special tool that he has for that purpose.

If you do not have the selvage removed by the dealer, you must remove it yourself. It is easiest to remove the selvage from the wallpaper rolls before cutting the rolls into strips. Then, when you start hanging wallpaper, this work is already out of the way. Remove selvage by the procedures on Page 15.

Measuring and Cutting Wallpaper – Removing Selvage

CAUTION

Be careful not to damage flock surfaces by rubbing or pressing hard with straightedge.

1. Unroll part of wallpaper.

To be sure that a hairline, visible edge of selvage does not remain, remove 1/16-inch of pattern along with the selvage.

2. Align straightedge with selvage [1].
3. Using sharp knife, cut off selvage [1].
4. Unroll wallpaper from which selvage has not been trimmed. Re-roll wallpaper from which selvage has been trimmed. Align straightedge and cut off selvage.
5. Repeat Step 4 until all selvage has been removed from roll. Repeat for all rolls.

Measuring and Cutting Wallpaper Strips for Ceilings

Wallpaper for ceilings frequently has a pattern that requires no matching. In this case, strips do not require extra length for adjusting patterns. Cut strips 2 inches longer than length required to go across ceiling. This extra length will allow 1-inch overlap onto walls at each end. The overlap will be covered when papering the walls.

If wallpaper has a pattern which requires matching, strips must be long enough to allow for adjustment of patterns and trimming. Allow 2 inches extra length at each end of strip.

1. Measure distance from wall to wall.
2. Add 2 inches or 4 inches as required to distance. Mark wallpaper.
3. Using straightedge and sharp knife, cut strip [1].
4. Determine number of same-length strips [2] required.
5. Match and cut remaining strips of same length.

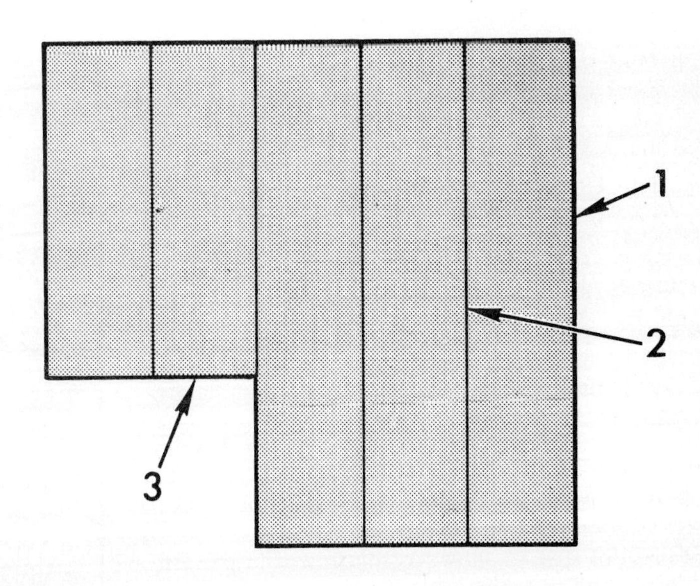

If ceiling has a complex shape which requires that other lengths be cut, try to make remaining strips [3] on ceiling run in same direction as strips [2].

6. Repeat Steps 1 through 5 for remaining strips [3] of other lengths.

▶ Measuring and Cutting Wallpaper Strips for Walls

Before cutting the first strip, determine where the pattern should end at the ceiling. This is a matter of pattern and personal preference. For example, a pattern with human figures should not end with the heads cut off or just the feet showing at the ceiling.

It is also important to remember that the line formed where the wall meets the ceiling is usually uneven. Because most walls are painted, this unevenness is seldom noticed. However, you may have wallpaper with a pattern that forms a strong horizontal line. If this strong horizontal line is located near the ceiling line, the ceiling can appear noticeably uneven and the wallpaper can look poorly matched. Therefore, locate strong horizontal lines in patterns as far as possible from the ceiling line.

Measuring and Cutting Wallpaper Strips for Walls

1. Unroll enough wallpaper to reach from ceiling to floor.
2. While holding wallpaper against wall, move paper up and down. Select preferred location for ending pattern at ceiling.
3. Mark wallpaper where strip ends at ceiling.
4. Mark wallpaper where strip ends at baseline.
5. Place wallpaper on pasting table or other flat surface. Two inches should be added to top of strip to allow for adjustment of pattern and trimming.
6. Using straightedge and sharp knife, cut wallpaper 2 inches above mark at top of strip.

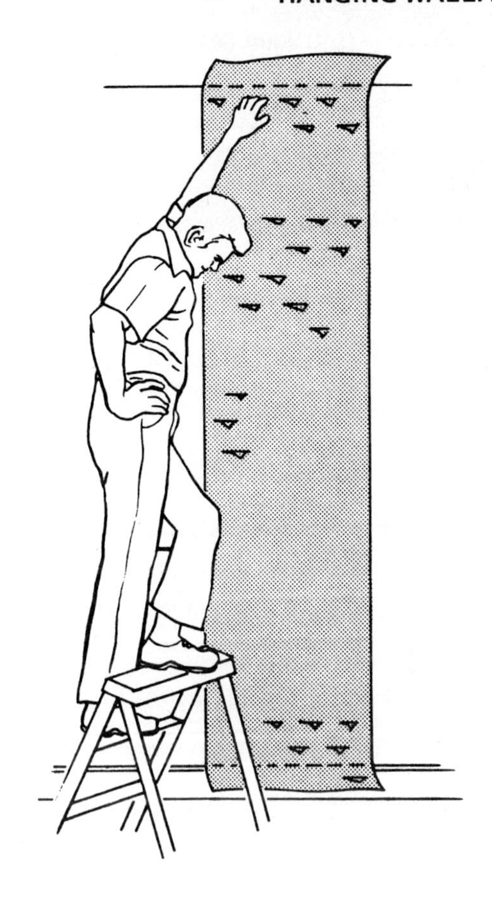

Measuring and Cutting Wallpaper Strips for Walls

7. Using straightedge and sharp knife, cut wallpaper 2 inches below mark at bottom of strip.

8. Holding strip in position against wall, check that

   - Pattern ends "correctly" at ceiling, and

   - Strip is long enough to permit adjustment and trimming.

In matching patterns, attempt to match strips in a way that results in least amount of waste. Different patterns result in different amounts of waste. If no matching is required, very little waste results. If you have an 8-foot wall and the pattern repeats itself in even fractions of 8 feet, you will have very little waste.

Measuring and Cutting Wallpaper Strips for Walls

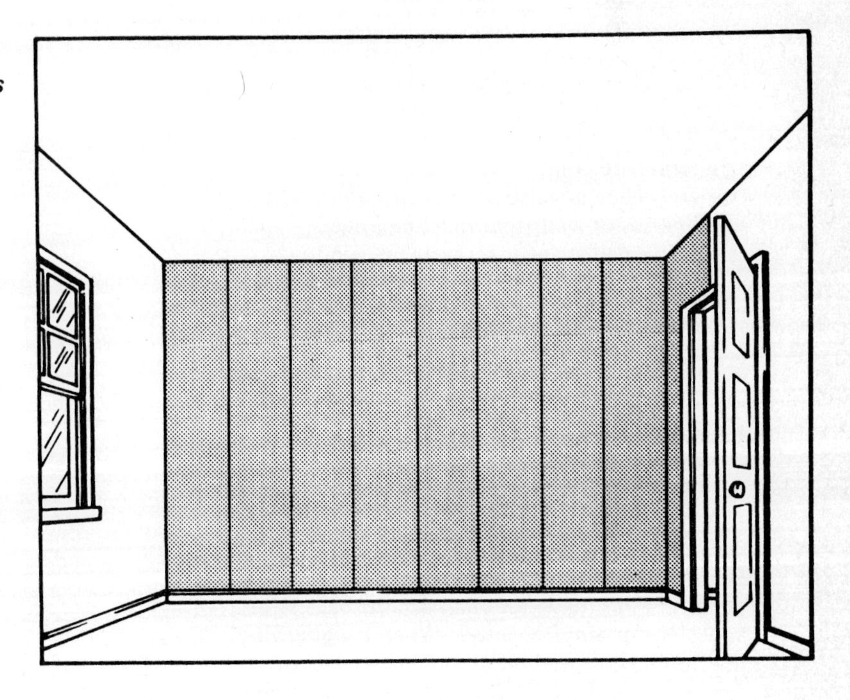

With some patterns, you should match each strip to the preceding strip. With other patterns, you should match alternate strips. To find the best match for your pattern, you must experiment.

9. Align wallpaper to first strip until it matches correctly with least amount of waste. Cut to same length as first strip.

Wall or room must be papered in sequence. You cannot skip strips for doors, windows or other obstacles and come back to them later.

10. Match and cut only enough full length strips to go from first strip to first obstacle.

Remaining full-length strips and pieces may be cut as required.

HANGING WALLPAPER

▶ Making Seams – Butt Joint Seams

Butt joints [1] have become almost the only method used for making seams between strips of wallpaper. They are by far the least noticeable of the joints.

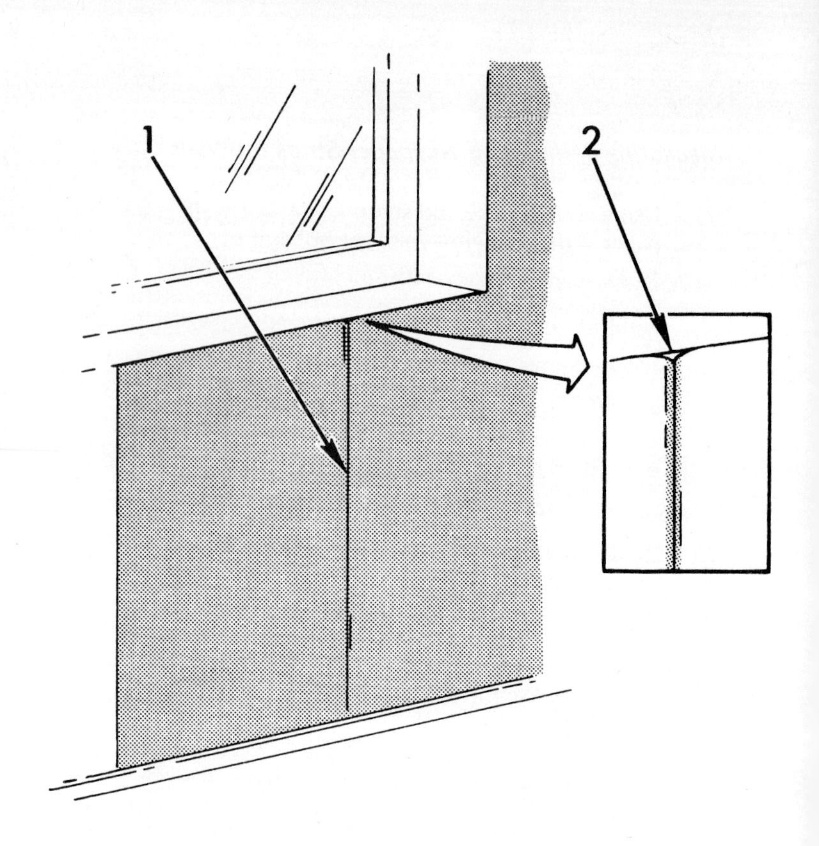

A butt joint [1] is made by moving the strip being hung tightly against the strip already hung.

Firmly slide the strip tightly against the preceding strip until a small ridge [2] rises at the seam. As the paste dries, the wallpaper will shrink, causing the ridge to flatten.

▶ Making Seams – Hairline Joint Seams

Hairline joints are used only for unpatterned backing paper. Backing paper is generally applied to the wall before hanging foil wallpapers. Because the paper will be covered by patterned wallpaper, the crack will not be seen.

A hairline joint is made by moving the strip being hung until its edge touches the edge of the strip already hung.

As the paste dries, the paper will shrink, causing a small hairline gap between the edges.

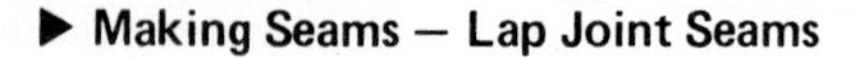

▶ Making Seams – Lap Joint Seams

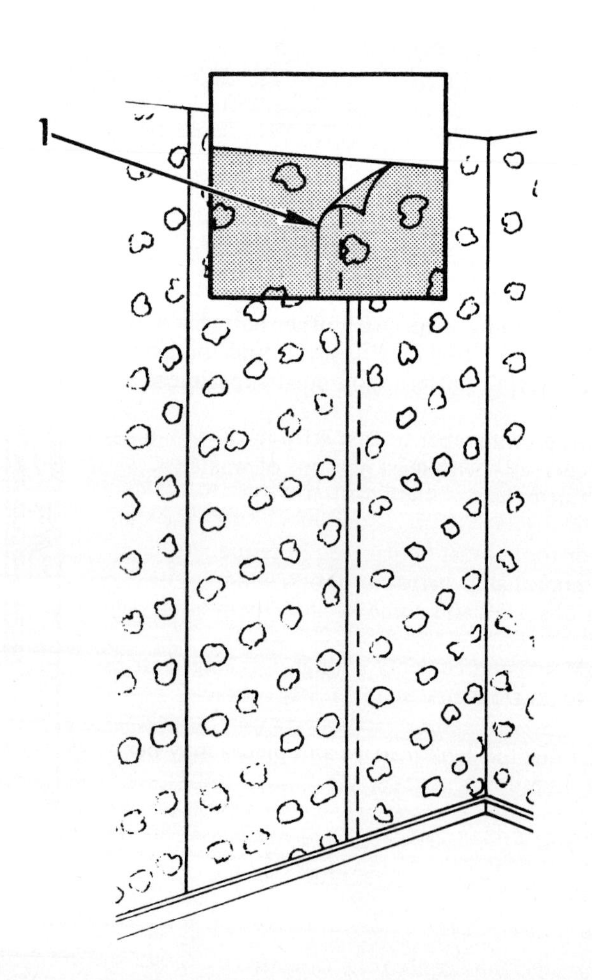

Lap joints [1] are seldom used to make seams between strips of wallpaper. They are noticeable and unattractive for that purpose.

Lap joints are commonly used to make seams at corners. They are also used in situations where small pieces of wallpaper must be joined – such as fitting small strips in a casement window.

In years past, lap joints were frequently used. Wallpaper ordinarily had an unprinted strip about one inch wide along each edge. This small strip, called selvage, was frequently left on one edge and covered with the adjoining wallpaper strip. Few wallpapers are now sold with selvage.

CAUTION

Do not use lap joints [1] when hanging vinyl wallpaper. Pastes made for vinyl do not have strong adhesion for vinyl on vinyl. When hanging vinyl, use double cut method of making wallpaper seams. **Go to Page 19 for making a double cut.**

▶ Making a Double Cut

A double cut is made at seams where the wallpaper strips have been overlapped. Lap joints are often necessary to match patterns when papering around windows and doors or into corners.

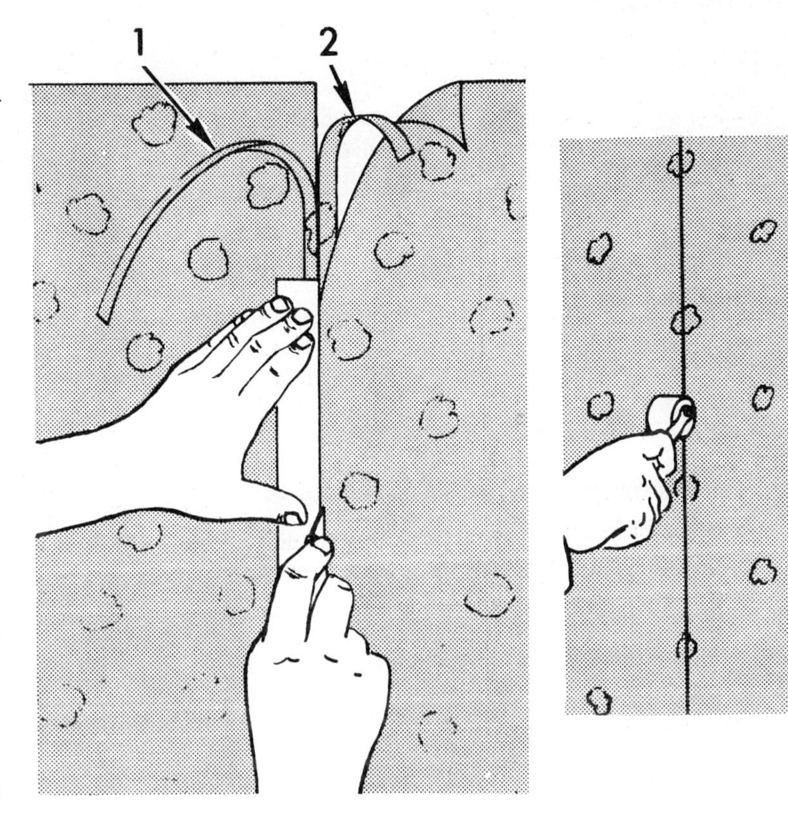

CAUTION

When vinyl wallpaper is hung, all lap joints must be double-cut to remove overlap. Vinyl adhesives do not offer strong adhesion for vinyl on vinyl.
For cutting wallpaper, cutting stroke must be firm enough to cut through both layers of wallpaper.

1. Place straightedge along center of overlap. Using one firm stroke with trimming knife, cut through both layers of wallpaper.

2. Carefully remove strip [1] from top wallpaper. Carefully pull top wallpaper from wall until strip [2] of bottom wallpaper can be removed. Carefully remove strip [2].

3. Using smoothing brush, smooth wallpaper against wall.

4. Using seam roller or damp cloth, firmly press edges of wallpaper to wall.

▶ Using a Seam Roller

In many cases, the edges [1] of wallpaper strips curl up away from the wall (or ceiling) as the wallpaper adhesive dries. A seam roller [2] is used to press the wallpaper edges firmly against the wall.

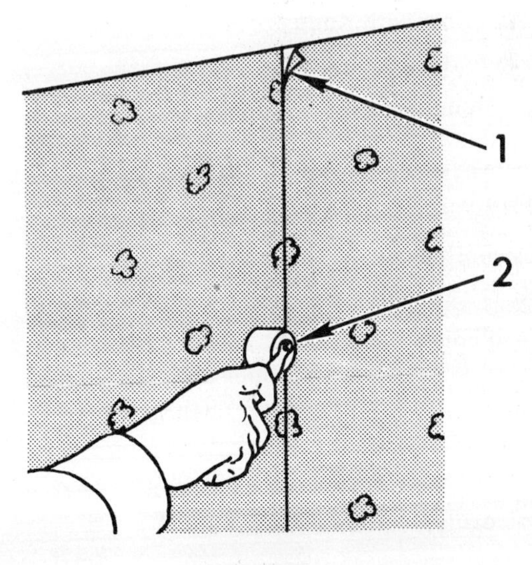

If the seam roller is used immediately after the strip is hung, the adhesive will be wet. The roller will simply force the wet adhesive out from under the edge. The edge will come loose from the wall.

Before using the seam roller, it is necessary to wait for the adhesive to begin to dry. Then when the edge is pressed with the seam roller, adhesive will not be forced out from under the edge. The edge will stick well to the wall.

The proper time for using the seam roller must be determined by trial and error. It should be possible to press firmly on the edge without squeezing out much adhesive. It may be possible to hang several strips before returning to the first seam to use the roller.

CAUTION

A seam roller will press down flocks and raised patterns. The flattened appearance will be noticeable. To avoid damage to the appearance of flocks and raised patterns, do not use a seam roller on them. Instead, press seams and edges firmly with a damp cloth.

HANGING WALLPAPER

▶ Trimming Wallpaper

For trimming wallpaper at ceilings and baseboards, a trimming wheel is the best tool to use. Trimming wheels are available with knife edges which score wallpaper or serrated edges which perforate wallpaper. With either type of trimming wheel, it is unlikely that wallpaper will be cut through completely. Therefore, a scissors will be required to complete the cutting.

For trimming wallpaper at windows, doors, and other obstacles, a knife is the best tool to use. When wallpaper is wet, a knife will frequently tear the wallpaper rather than cut it. A scissors is often required to complete the cutting.

▶ Trimming Wallpaper with Trimming Wheel

Trimming wheel is rolled across wallpaper along corner [1], where excess paper is to be trimmed. Trimming wheel makes a perforation or score along lines.

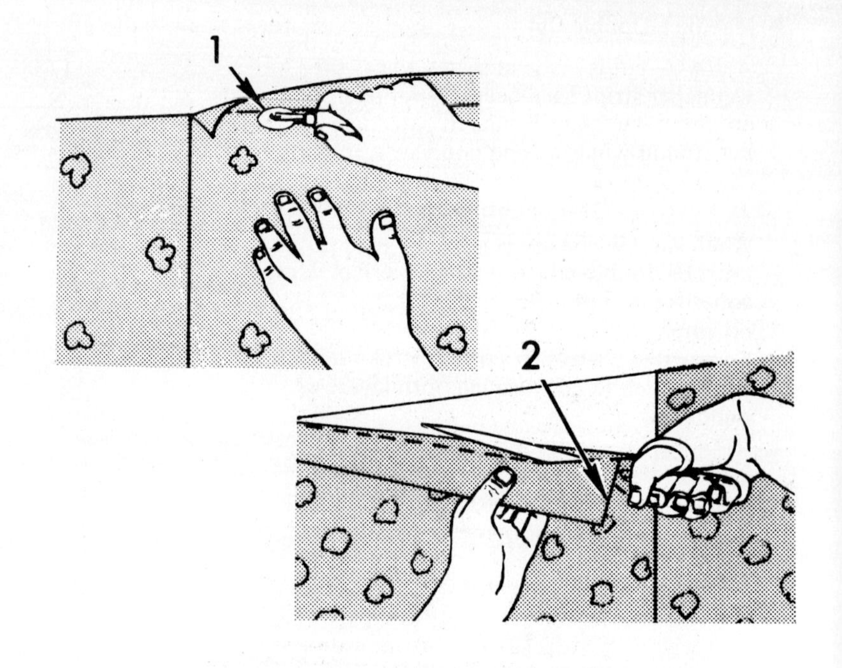

1. Using smoothing brush, firmly pat paper into corner.
2. Firmly pushing trimming wheel, make perforation or score along corner [1], where excess paper is to be trimmed.
3. Carefully pull wallpaper [2] away from wall. Using scissors, cut and remove excess paper along perforation or score.
4. Using smoothing brush, smooth wallpaper against wall.

▶ Trimming Wallpaper with Knife

In trimming excess paper with sharp knife, a straightedge should be used to ensure a straight cut. If straightedge cannot be used, edge of obstacle where excess paper is to be trimmed should be used to guide knife.

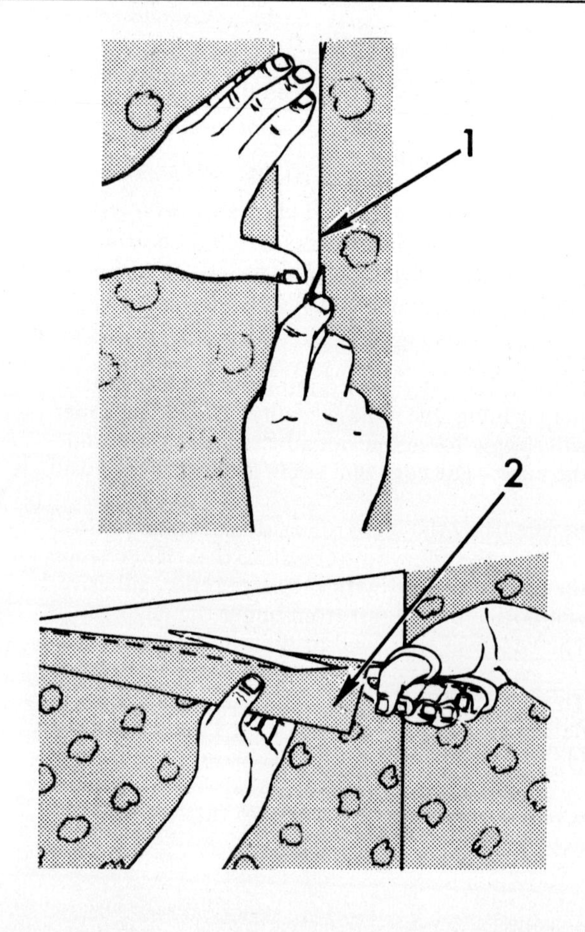

1. Using smoothing brush, firmly pat paper into corner.

2. Place and hold straightedge [1] along corner where excess paper is to be trimmed.

<u>CAUTION</u>

Knife may tear wet paper instead of cutting it. If paper tears easily, score paper [2] and cut with scissors.

3. Using straightedge [1] as a guide, cut and remove excess paper with trimming knife. Remove excess paper.

4. Using smoothing brush, smooth wallpaper against wall.

▶ Mixing Paste

Wallpaper pastes are available in either premixed or dry form. If using premixed paste, go to Page 22 for applying paste.

If using dry paste, continue.

For hanging nonporous wallpaper such as vinyl or foil, mildew resistant paste must be used. Use pastes such as vinyl adhesives which are not subject to mildew. Label on package will state whether paste is mildew resistant.

Mildew resistant additives can be added to pastes which are not ordinarily mildew resistant. Follow manufacturer's instructions for using mildew resistant additives.

For hanging heavy wallpaper, paste strengtheners should be added to paste. Follow manufacturer's instructions for using paste strengtheners.

Mixing Paste

For easier application of paste and better adhesion of wallpaper to surface, mix paste one hour before using.

Always read manufacturer's instructions before mixing paste. Use amount of water and paste specified by manufacturer.

1. Pour specified amount of water into clean bucket or pail.

Paste should be added to water in small amounts All lumps must be broken up. Paste should be smooth, lump free, and about as thick as a milkshake.

2. While stirring mixture by hand, add dry paste to water until mixture has correct thickness.

HANGING WALLPAPER

▶ Applying Paste

Read entire procedure before starting.

If the strip has been measured and cut so that the top and bottom 1 or 2 inches of the strip are to be removed during trimming, do not apply paste to them. If the ends are not pasted, it will be easier to keep the floor and ceiling clean and to unfold the strips during hanging.

1. Fill bucket, pail, or roller tray one-half full with paste.
2. Position strip so that edge [1] extends slightly past edge of table. Bottom end [2] of strip should be on your right. End [2] should extend slightly past edge of table.
3. Fill brush or roller with paste. Beginning at right end [2] and working to left, spread paste along center of strip until one-half length is covered.
4. Working from center of strip, spread paste to edge [1], until one-half length is covered.
5. Position strip so that edge [3] extends slightly past edge of table.
6. Working from center of strip, spread paste to edge [3] until one-half length is covered.

After length of wallpaper is pasted, it is folded to make it easier to handle while hanging it and to allow wallpaper to absorb paste thoroughly and evenly.

To make it easier to distinguish the top end of strip from the bottom end, make top fold slightly longer than bottom fold.

CAUTION

Be careful not to crease wallpaper when folding. If folding foil wallpaper be especially careful to avoid creases. Creases cannot be removed from foil.

7. Fold bottom end [4] less than one-quarter distance toward middle of strip.
8. Pull strip toward right edge of table until remainder of strip is on table. Edges [5,6] should extend slightly past edges of table.
9. Starting at center and working toward edges [5,6,7], apply paste to remainder of strip. Note position of strip on table in illustration.
10. Fold top end [2] of strip more than one-quarter distance toward middle of strip.

Paste on strip should cure about 10 minutes before strip is hung.

11. Remove strip from table. Place strip in clean area while paste cures.

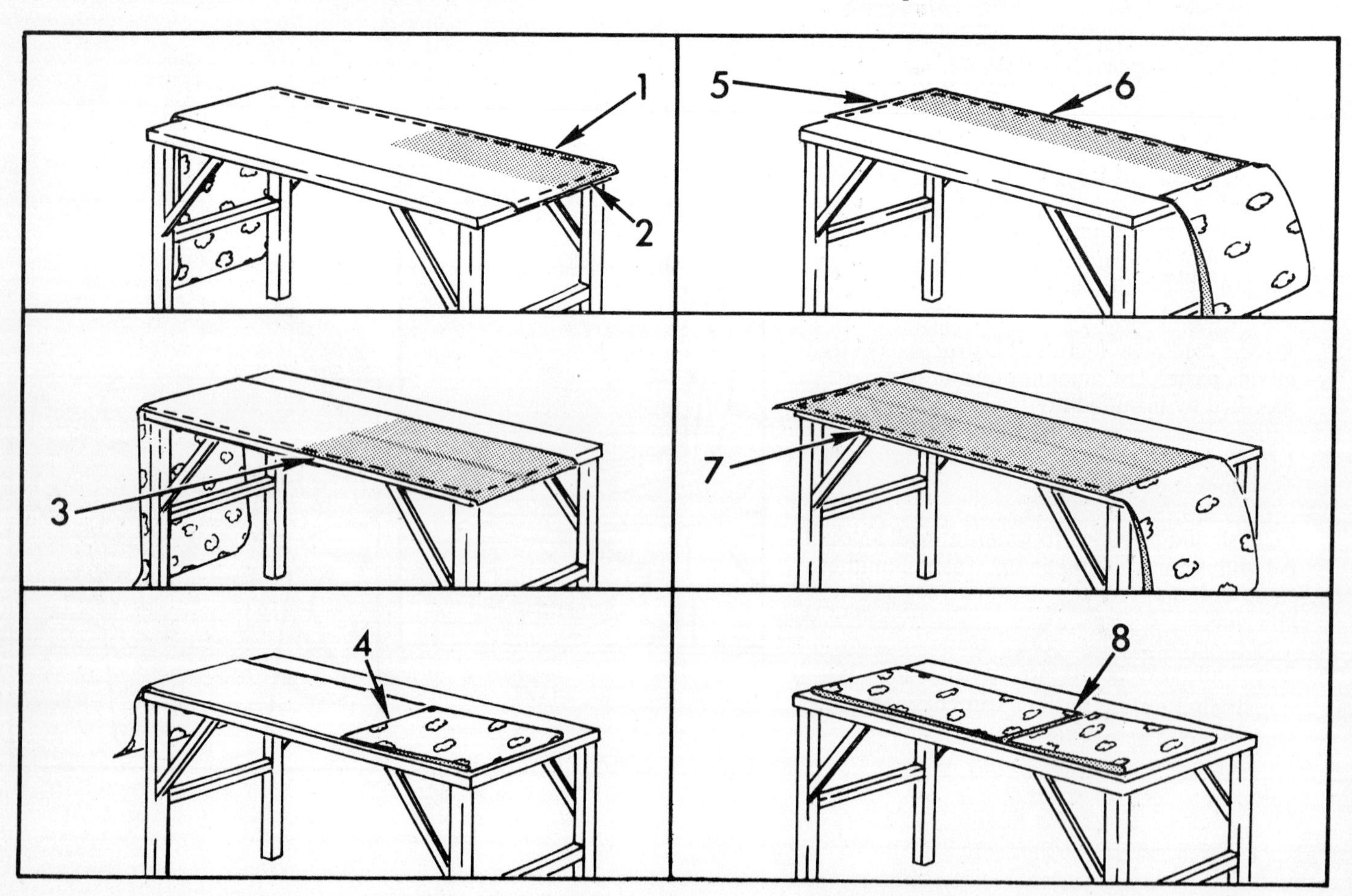

▶ Papering Ceilings

If you are papering the ceiling and walls, complete papering of ceiling before starting papering of walls.

Papering a ceiling is a difficult operation for an amateur. Although it can be done without a helper, the job will be accomplished much easier with an assistant to help handle paper and tools.

The following tools and supplies are required:

- Tools and supplies for hanging wallpaper. Pages 9 and 10.
- Scaffold [1] constructed of plank and two stepladders

Ceiling surfaces must be prepared. Page 7.
Chalk line for first strip must be marked. Page 14.

First strip and all adjoining strips of same length should be cut. Page 15.

Ceiling fixtures which interfere with papering should be removed. Page 37.

Read entire procedure before starting to paper. Also be sure to read Making Seams, Page 18, Using a Seam Roller, Page 19, and Trimming Wallpaper, Page 20.

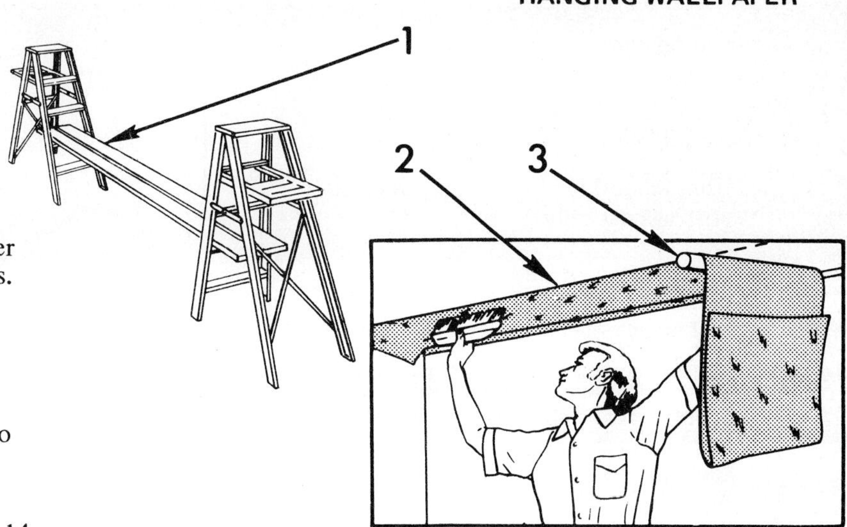

1. Apply paste to strip. Page 22.

2. Unfold starting end [2] of strip.

Smoothing brush should be placed about 2 feet from starting end [2] of strip.

Broomstick or other long, rounded object [3] which will not tear wallpaper should be placed in position to support remainder of strip.

Papering Ceilings

3. While holding strip near ceiling, align edge with chalk line [2] or edge of preceding strip. Check that starting end [1] of strip will overlap wall as required.

4. Using smoothing brush, press enough wallpaper against ceiling to hold end of strip in place.

Wallpaper can be repeatedly applied to and removed from a surface without damage to paper. Wallpaper paste dries slowly (10, 20 or more minutes) so there is no need to hurry during handling.

Rough alignment of wallpaper is made by pulling it from ceiling and reapplying it. Final alignment is made by striking it firmly in desired direction with smoothing brush or by pressing against it with palms of hands and pushing firmly in desired direction.

5. Check alignment of strip. Realign strip as required. When it looks as if strip is being started straight, go to next Step.

6. Using smoothing brush, smooth center [3] of strip against ceiling for a distance of about 3 feet.

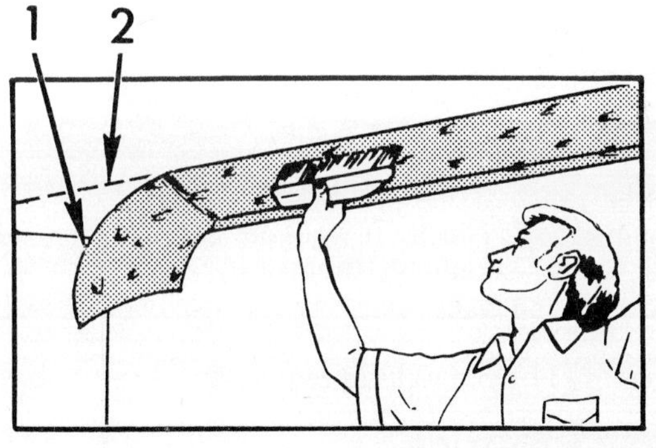

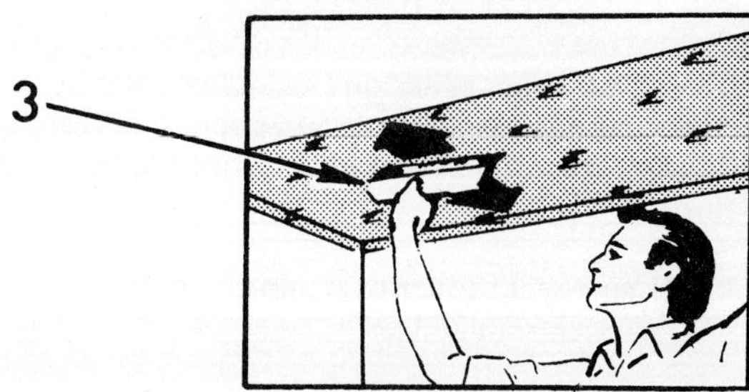

7. Working from center [3] of strip to edges, smooth remainder of 3-foot length against ceiling.
8. Check alignment of strip and realign as required.

9. Repeat Steps 6, 7 and 8 until entire strip is against ceiling.

HANGING WALLPAPER

Papering Ceilings

When smoothing strip against ceiling, be sure strip stays aligned.

10. Using smoothing brush, smooth entire strip against ceiling.

11. Using bristles of smoothing brush, gently tap wallpaper into corners [1] where ceiling and wall meet.

12. Check that there are no bubbles or wrinkles in entire strip. Remove any bubbles or wrinkles with smoothing brush. It may be necessary to cut bubbles to release trapped air.

First and last strips hung must have corners cut and pressed. If hanging first or last strip, go to Step 13. Last strip must be measured, cut and pasted before you perform Step 13. Be sure to allow for 1-inch overlap onto adjoining walls.

If hanging strips other than first and last strip, go to Step 16.

13. Using bristles of smoothing brush, tap wallpaper into corner [2]. Place mark on wallpaper at exact corner position.

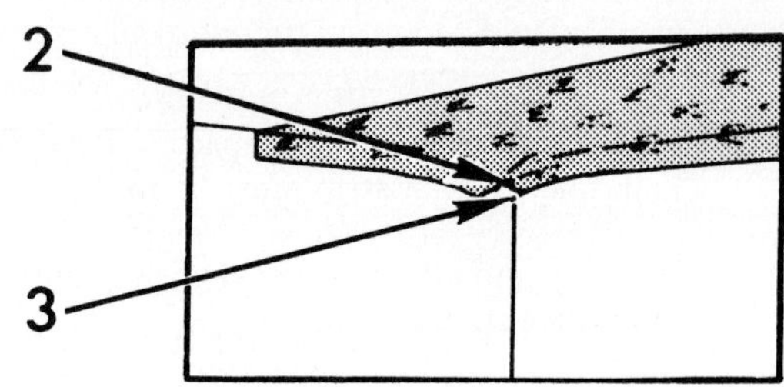

14. Gently pull about 1 foot of wallpaper from ceiling at corner [2]. Using scissors, cut a square [3] from wallpaper, using mark as a guide.

15. Smooth wallpaper into corner by tapping gently with smoothing brush.

Papering Ceilings

If walls are to be papered, overlap [1] onto walls should be trimmed to 1/2-inch. If walls are not to be papered, overlap should be trimmed to edges where ceiling and walls meet.

16. Trim overlap [1] from walls as required.

CAUTION

Do not use seam roller on flocked or raised-pattern wallpaper. Damage to paper will result. Instead, press seams and edges firmly with damp cloth.

17. Using seam roller or damp cloth, firmly press all seams and edges.

18. Using moist sponge, remove paste from wallpaper and walls. Rinse sponge frequently with clean water.

19. Measure, cut, paste and hang remaining strips.

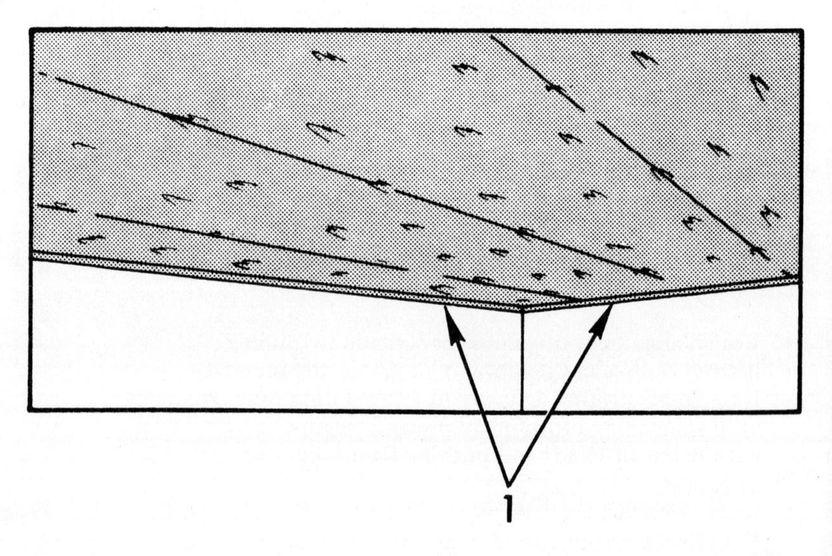

▶ Papering Walls

Instructions in this section show how to apply wallpaper to walls. Read entire procedure before starting.

Before this section can be performed:

- Wall surfaces must be prepared. Page 7.
- Plumb line for first strip must be marked. Page 11.
- First strip and adjoining strips of same length should be cut. Page 16.
- Fixtures which interfere with papering should be removed. Page 36 and Page 37.

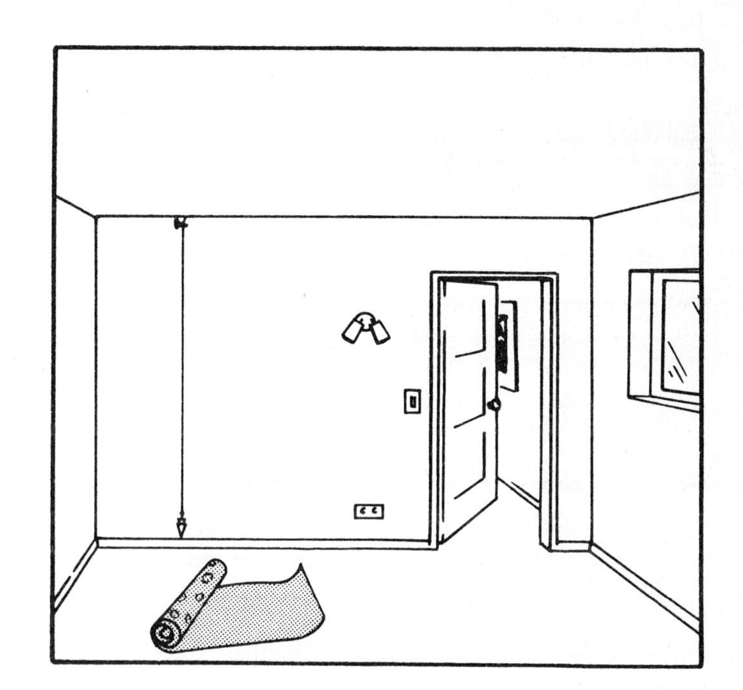

Before hanging wallpaper around an obstacle, read the following sections as applicable:

- Using a Seam Roller, Page 19.
- Papering Around Corners, Page 28.
- Papering Around Windows and Doors, Page 29.
- Papering Casement Windows, Page 31.

Papering Walls

If not hanging pre-pasted wallpaper, go to Step 3.

If hanging pre-pasted wallpaper, continue.

1. Fill water tray [3] one-half full with tepid water. Place tray alongside wall at place where strip is to be hung.

CAUTION

When wetting pre-pasted wallpaper, do not soak longer than manufacturer recommends. Excess soaking may weaken paper or cause paste to fail.

2. Beginning at bottom of strip, roll strip loosely. Place rolled strip in water tray. Go to Step 5. Be sure to read note before Step 5.
3. Apply paste to strip. Page 22.
4. Unfold top end of strip.

First strip of wallpaper to be hung is aligned with chalk line [2]. All other strips are aligned with preceding strip. When aligning with strip, be sure to match patterns. Also, try to align pattern with ceiling line.

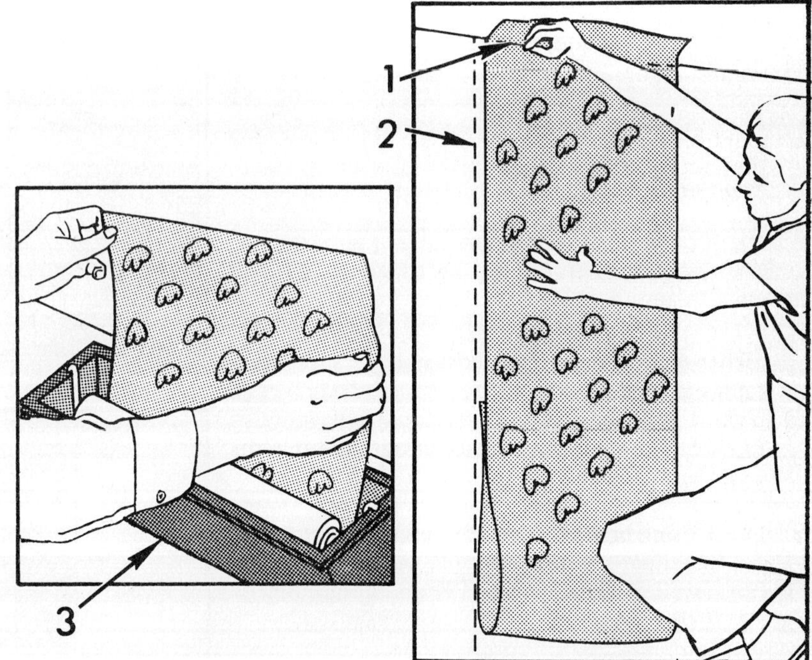

5. Place top end of strip against wall. Adjust strip until pattern crosses ceiling line [1] at place determined and edge of strip is aligned with chalk line [2] or preceding strip.

Papering Walls

6. While holding strip against wall, press about 1 foot of end of strip against wall with smoothing brush to hold strip in place.

Wallpaper can be repeatedly applied to and removed from the wall without damage to paper. Wallpaper paste dries slowly (10, 20 or more minutes) so there is no need to hurry during handling.

Rough alignment of wallpaper is made by pulling it from wall and reapplying it.

Final alignment is made by striking it firmly in desired direction with smoothing brush or by pressing against it with palms of hands and pushing firmly in desired direction.

7. Check that strip is aligned and patterns are matched. If not aligned, align strip and match patterns.

8. Unfold bottom half of strip, if required.

When smoothing strip against wall, always smooth down about 3 feet at a time, starting at top. Smooth center of strip before smoothing edges.

9. Using smoothing brush, smooth center of strip against wall for a distance of about 3 feet.

10. Working from center of strip to edges, smooth wallpaper against wall.

11. Repeat Steps 9 and 10 until entire strip is smoothed against wall.

Papering Walls

12. Check that strip is aligned and patterns are matched. If not aligned, align strip and match patterns.

When smoothing strip against wall, be sure strip stays aligned.

13. Using smoothing brush, smooth entire strip against wall.

14. Using bristles of smoothing brush, gently tap wallpaper into corner [1] where ceiling and wall meet, at corners [2] and at molding or baseboards [3].

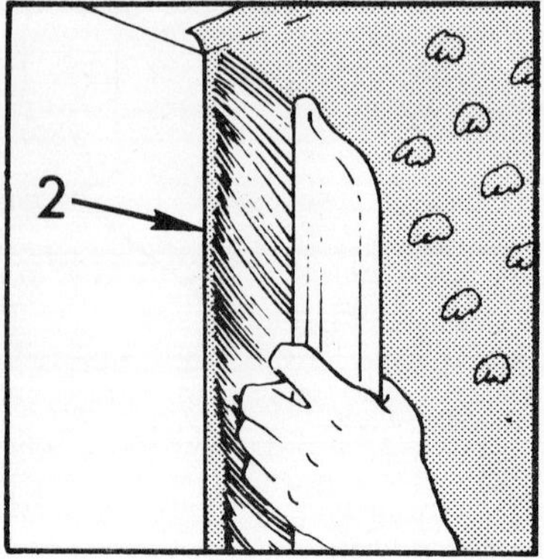

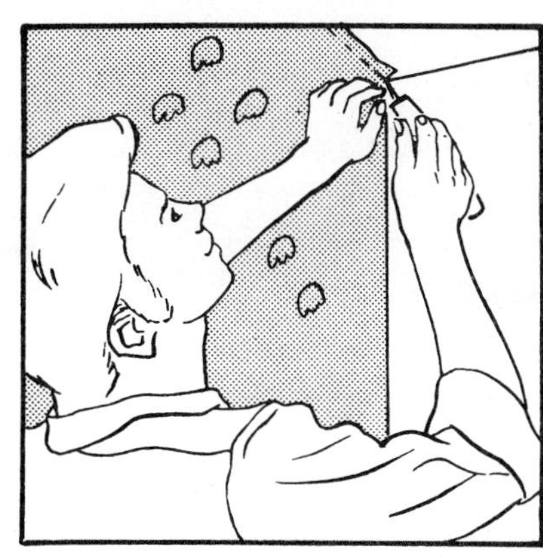

Papering Walls

15. Check that there are no bubbles or wrinkles in entire strip. Remove bubbles or wrinkles with smoothing brush. It may be necessary to cut bubbles to release trapped air.

16. Trim excess wallpaper at groove where ceiling and wall meet, at corners and at moldings or baseboards.

17. Using moist sponge, remove paste from wallpaper. Rinse sponge frequently with clean water.

Papering Walls

18. Using sponge and clean water, remove paste from walls, woodwork, and ceiling.

CAUTION

Do not use seam roller on flocked or raised-pattern wallpaper. Damage to paper will result. Instead, press seams and edges firmly with damp cloth.

19. Using seam roller or damp cloth, firmly press all seams and edges against wall.

20. Using sponge, carefully remove any paste squeezed from seam.

21. Measure, cut, paste and hang remaining strips.

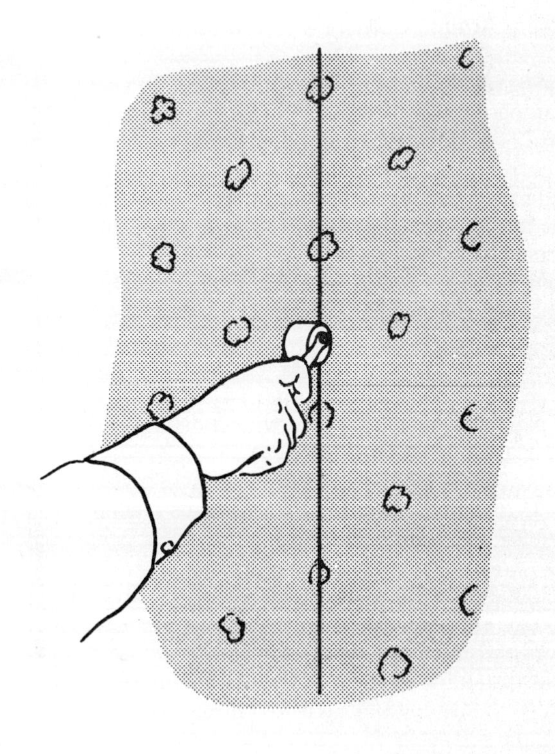

HANGING WALLPAPER

▶ **Papering Around Corners**

Instructions in this section tell how to hang wallpaper around outside corners and inside corners. If papering around an inside corner, go to bottom half of page.

▶ **Papering Around Corners – Outside Corners**

Special care should be taken when papering around an outside corner [1]. Do not rub on paper where it passes over the corner. The surface of the wallpaper could be rubbed off or damaged.

Outside corners are more noticeable than inside corners [5]. Mismatches between patterns are more obvious at outside corners [1]. Seams at or near the corner should be avoided if possible.

For papering around an outside corner [1], strip [2] is aligned with preceding strip [3] and folded around outside corner. Cut slit [4] in top and bottom ends of strip at corner before folding strip around corner.

Strip [2] is then trimmed. As the paste dries, the strip will be pulled tightly against wall.

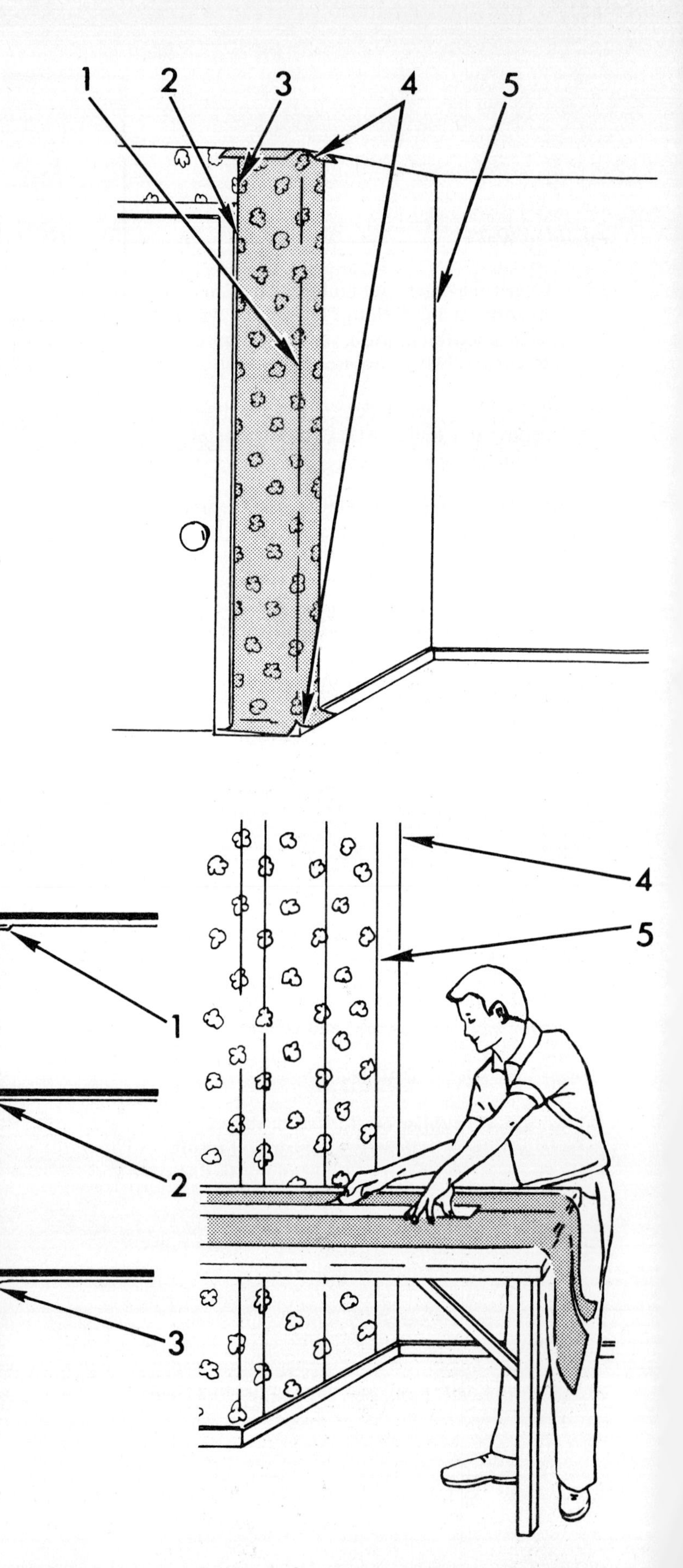

▶ **Papering Around Corners – Inside Corners**

Papering inside corners [4] often results in some mismatch between strips. If strips are carefully aligned, any mismatch will be less noticeable. The correct way to paper an inside corner is to cut the strip into two parts and overlap [1] them at the corner. Never simply fold a strip [2] and then paste it into a corner. It may pull away from the corner [3] when it dries, becoming unsightly and subject to damage.

1. Measure and record distance between preceding strip [5] and corner [4] at three places. Add 1/2-inch to widest distance measured.

2. At distance figured in Step 1, place several marks on unpasted strip.

3. Using marks as a guide, cut strip with straightedge and trimming knife.

Two strips [6,7] are now available. Strip [6] cut to width determined in Step 1, is hung first.

Papering Around Corners – Inside Corners

4. Paste, hang, and trim first strip [1].

5. Measure and record width of second strip [5]. Beginning at corner [1], measure recorded distance. Place mark [2] on wall at recorded distance.

6. Place plumb line at mark [2]. Go to Page 11 for making a plumb line.

Second strip [5] is aligned with plumb line [4] on wall. Seam at inside corner [3] may have some mismatch after strip is hung.

7. Paste, hang, and trim second strip [5].

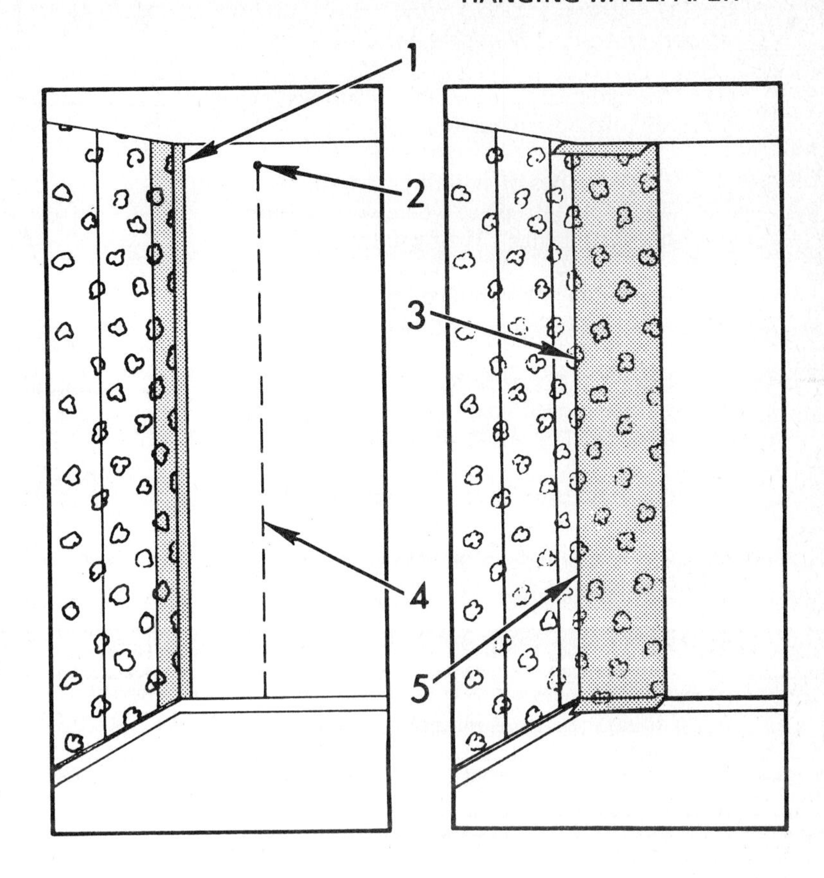

CAUTION

When vinyl wallpaper is hung, any lap joint at corner should be double-cut. Vinyl adhesives do not offer good adhesion for vinyl on vinyl.
Go to Page 19 for making a double-cut.

▶ Papering Around Windows and Doors

Instructions in this section tell how to cut and hang wallpaper around a window or door.

Procedures are provided for papering around a window only. However, papering around a door is identical to papering around a window except that cutting the wallpaper around the bottom edge is not necessary.

The first strip [1] to be hung around a window is aligned with the preceding strip [2].

The strip [1] is then pressed against the wall until it reaches the window.

The top and bottom edges of the strip [1] are not pressed against the wall until all cuts around the window are made.

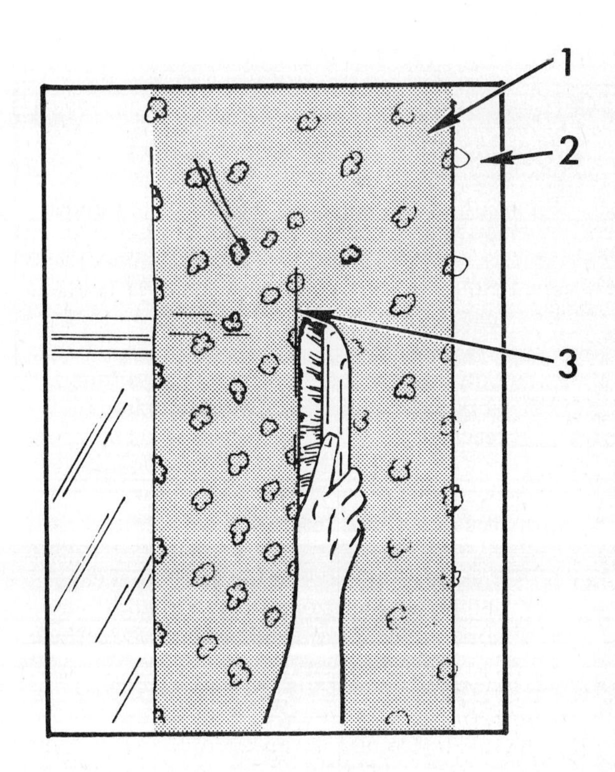

1. Align strip [1] and match patterns with preceding strip [2]. Press strip against wall until it reaches window.

2. Using bristles of smoothing brush, gently tap wallpaper into place where wall and vertical edge [3] of window frame meet.

Papering Around Windows and Doors

3. Beginning at top corner [1], cut wallpaper at 45° angle with scissors.

4. Using bristles of smoothing brush, gently tap wallpaper into place where wall and top horizontal edge [2] of window frame meet.

5. Beginning at bottom corner [3], cut wallpaper at 45° angle with scissors.

6. Using bristles of smoothing brush, gently tap wallpaper into place where wall and bottom horizontal edge [4] of window frame meet.

7. Press and smooth strip against wall above window. Press and smooth strip against wall below window. Trim excess paper at ceiling line and baseboard.

8. Trim excess paper from around window frames. Remove all excess paste.

If less than one strip is required to complete papering around window, go to Step 11.

If more than one strip is required to complete papering around window, go to Step 9.

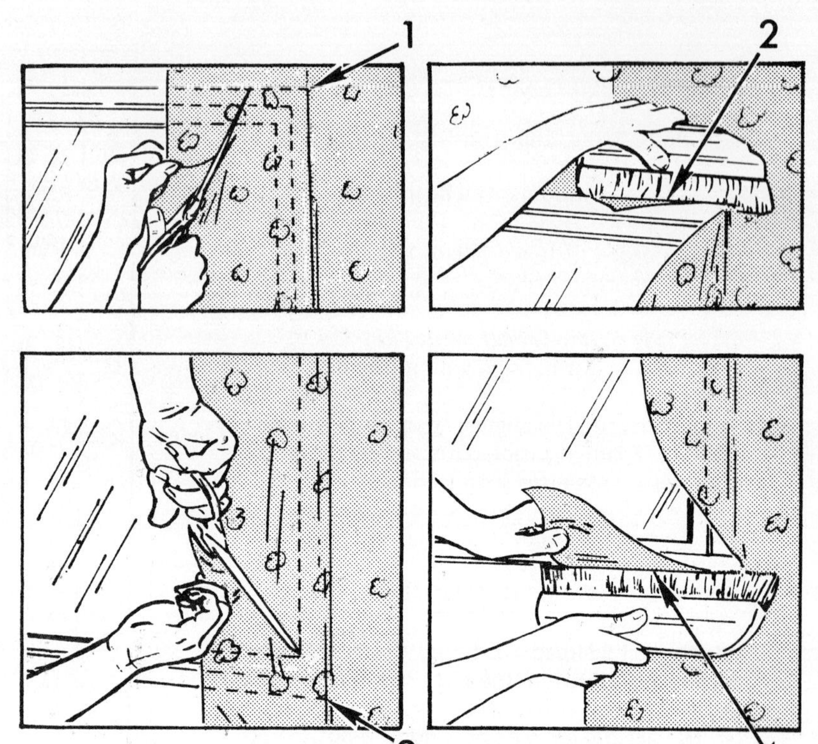

Papering Around Windows and Doors

9. Cut, paste, hang, and trim strips [1] above window.

10. Cut, paste, hang, and trim strip [2] below window.

11. Cut, paste, and hang strip [3] until top horizontal edge [4] of window frame is reached.

12. Beginning at top corner [5], cut wallpaper at 45° angle with scissors.

13. Press section between top horizontal edge and bottom horizontal edge until bottom horizontal edge of window frame is reached.

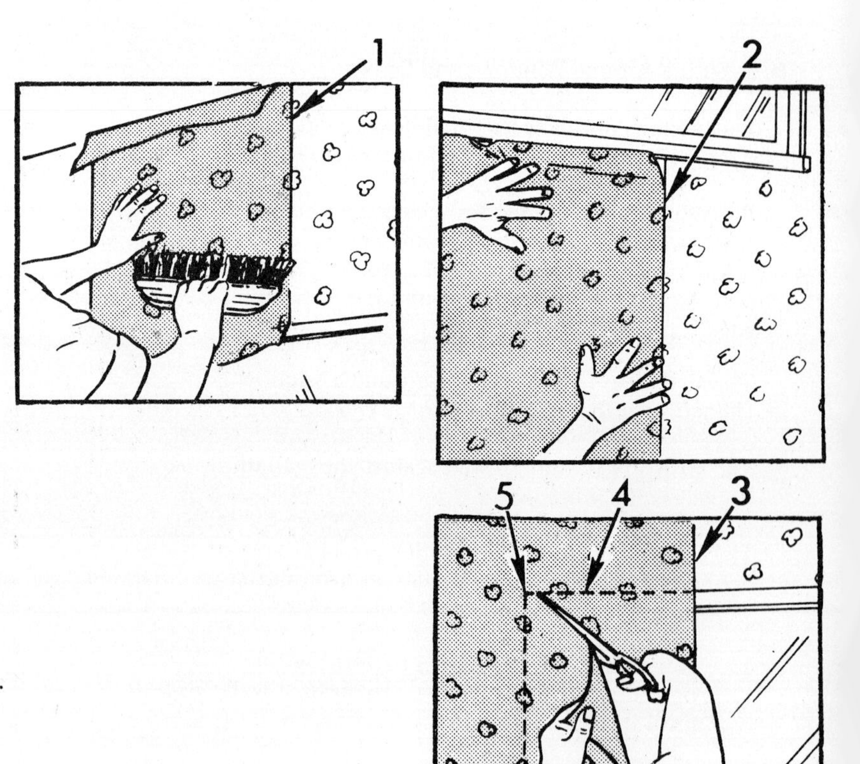

Papering Around Windows and Doors

14. Beginning at bottom corner [1], cut wallpaper at 45° angle with scissors.

15. Align, press, and smooth entire strip against wall.

16. Using bristles of smoothing brush, gently tap wallpaper into place where top horizontal edge [2], bottom horizontal edge, and vertical edge [3] of window frame meet wall.

17. Trim excess paper from window frame [4]. Remove all excess paste.

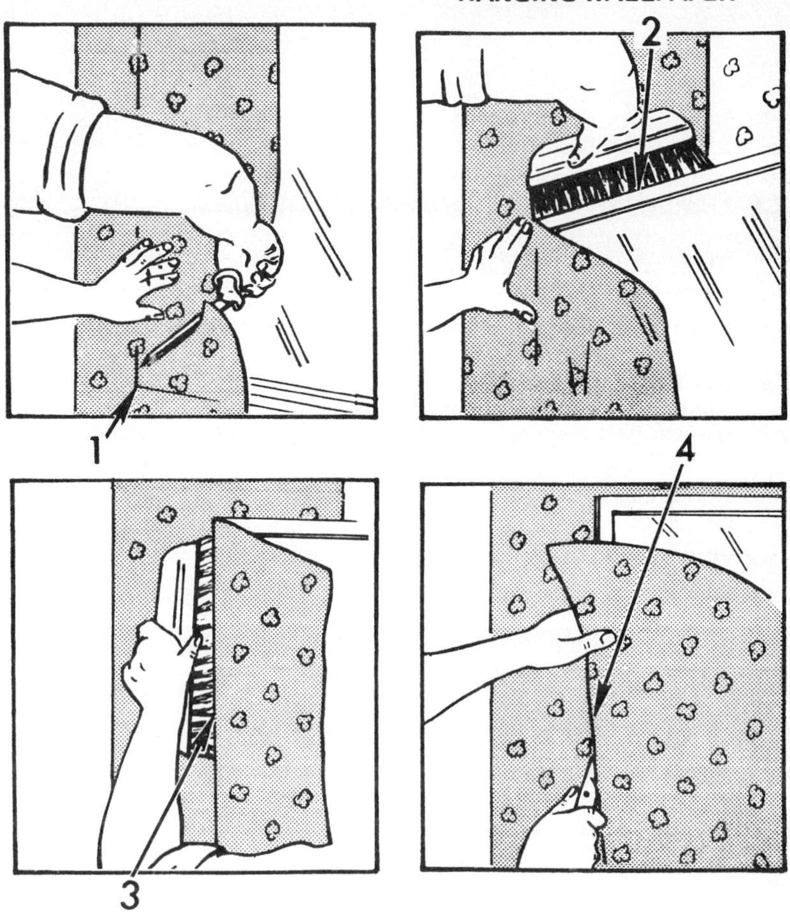

▶ Papering Casement Windows

Professionals generally paper the casements of casement windows. The wallpapering job has a much better appearance if the casements are papered. Sometimes the lower casement [1] is painted rather than papered because, since it is also a ledge, it is subject to a lot of wear and tear.

When papering a casement window, you should paper only the casements. The metal frames [2] which hold the glass panes are not papered. They are painted or left natural.

It is easy to paper casement windows. The key to getting professional-looking results is making the correct cuts and seams at the corners.

There are several different procedures for making the correct cuts and seams. One method is used for vinyl wallpapers; the other method is used for all wallpaper other than vinyl.

If you are papering your casement windows with wallpaper other than vinyl, go to Page 35.

If you are papering your casement windows with vinyl wallpaper, go to Page 32.

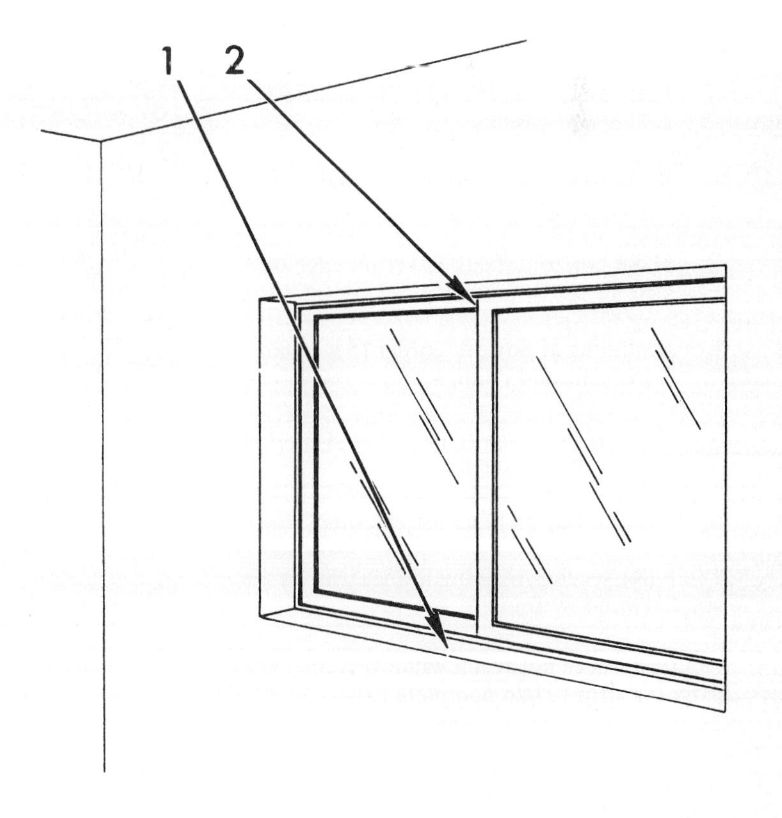

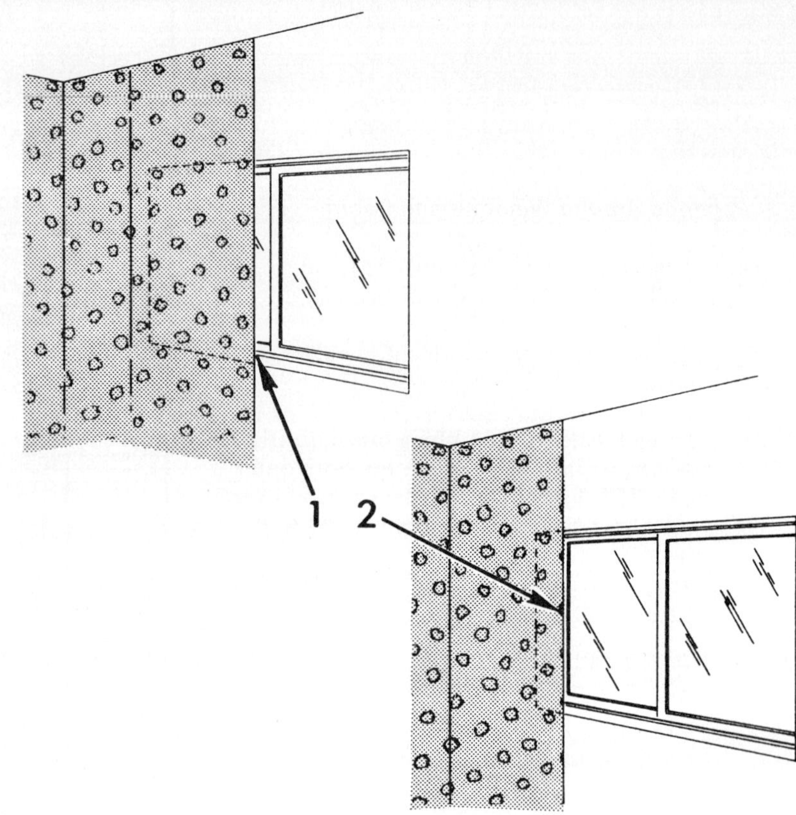

▶ Papering Casement Windows with Vinyl Wallpaper

There are two procedures for papering with vinyl wallpaper. These two procedures are given so that you can select the procedure which results in the least amount of waste. The method which you should use depends upon the amount of overlap between the window and the wallpaper strip.

If over half the strip [1] overlaps the window, go to Page 34 for cutting and trimming procedures for long overlap.

If less than half the strip [2] overlaps the window, go to next section (below) for cutting and trimming procedures for short overlap.

▶ Papering Casement Windows with Vinyl Wallpaper – Short Overlap

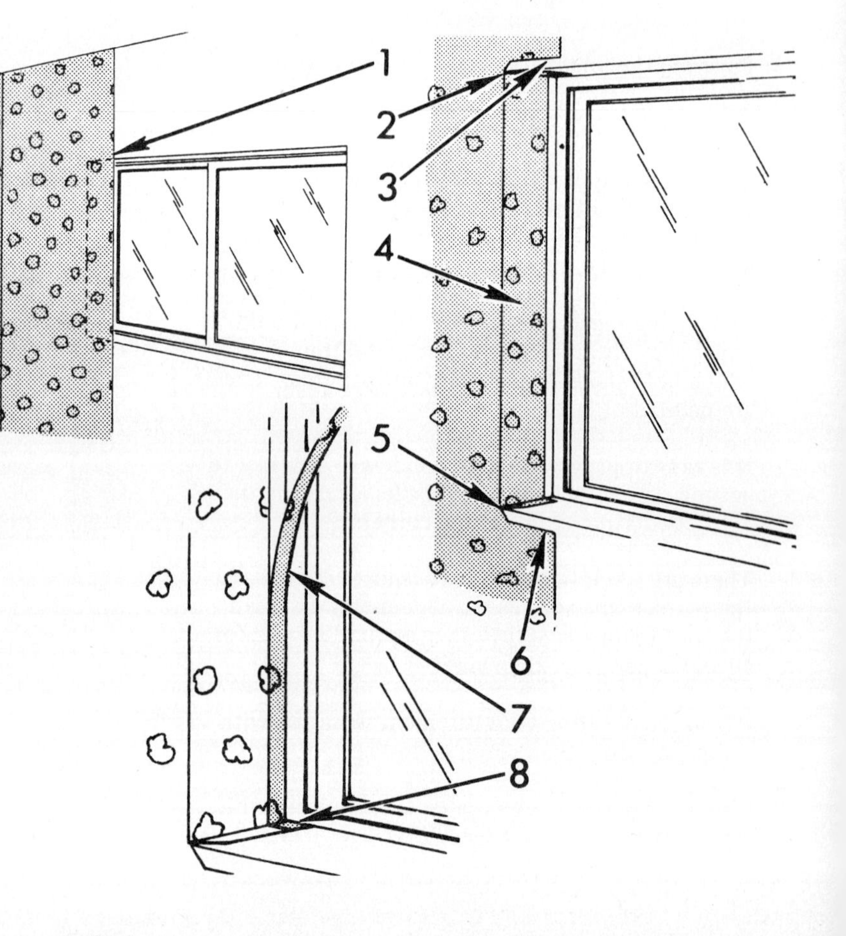

1. Paste and hang strip [1]. Trim strip at ceiling and baseboard.

2. Beginning at top corner [2], make 1-inch slit at 45° angle.

3. Make horizontal cut [3] from edge of strip to top end of slit.

4. Beginning at bottom corner [5], make 1-inch slit at 45° angle.

5. Make horizontal cut [6] from edge of strip to bottom end of slit.

6. Smooth flap [4] against casement.

If flap [4] reaches window frame, trim excess paper at window frame.

If flap [4] does not reach window frame, cut a matching strip [7] to complete papering of casement. Cut strip long enough to overlap top and bottom [8] corners. Trim strip [7] even with flap [4].

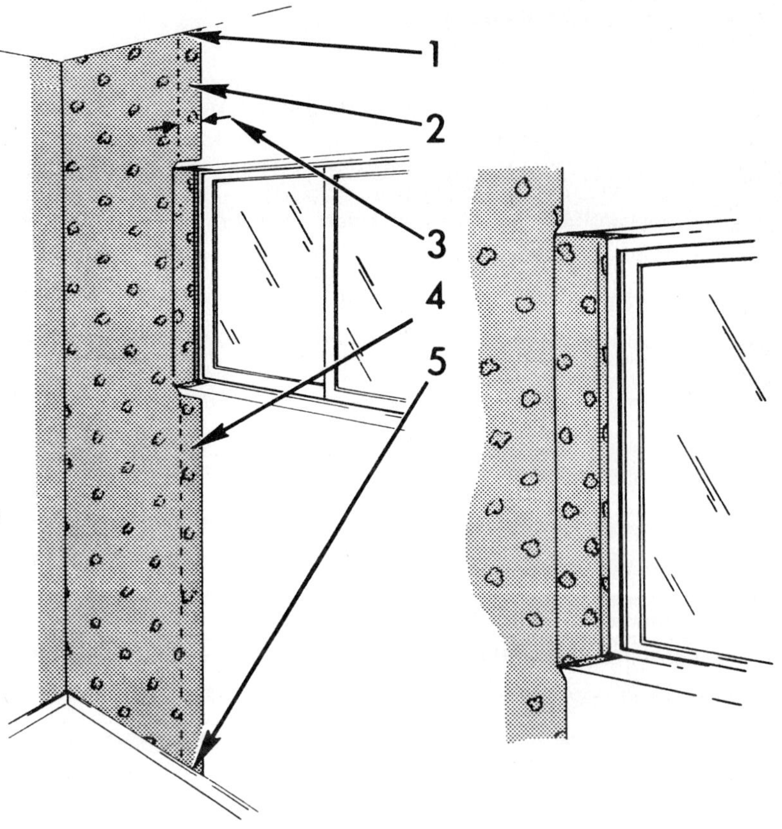

Papering Casement Windows with Vinyl Wallpaper – Short Overlap

Pieces [2,4] of strip must be cut and removed.

7. Measure distance [3] from edge of strip to 45° cut. Mark this distance on strip at ceiling [1] and baseboard [5].

8. Using straightedge and knife, cut and remove pieces [2,4].

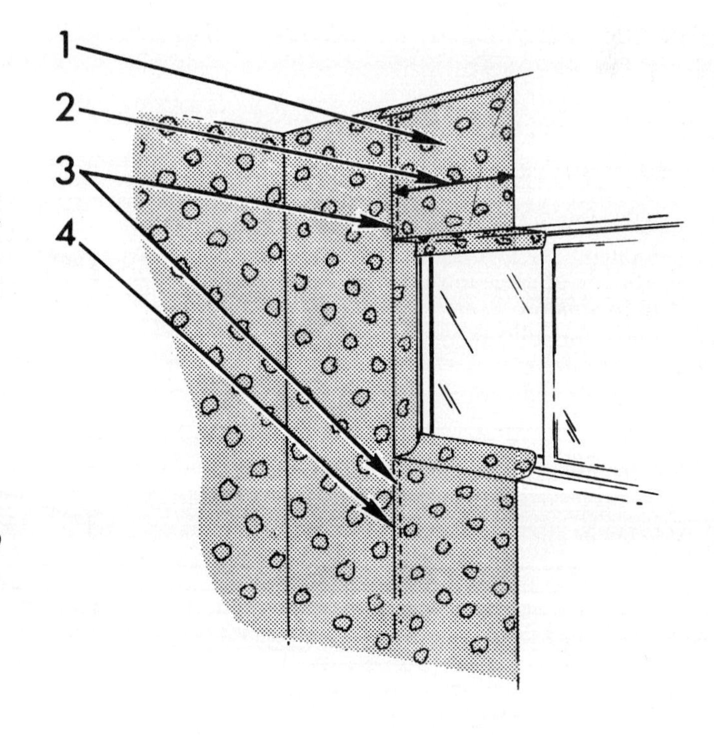

Papering Casement Windows with Vinyl Wallpaper – Short Overlap

9. From a new strip of wallpaper, measure and cut a matching piece [1] of wallpaper. Width [2] of piece [1] must be from edge of strip to edge of window casement. Length of piece must be from ceiling to window frame. Add 1 inch to length to allow for adjustment and trimming.

10. Paste, hang and trim piece [1].

11. Using procedures in Step 9, cut, paste, hang, and trim piece [4] from window to baseboard.

12. Double cut overlap joints [3]. Go to Page 19 for making a double cut.

13. Match, cut, and hang remaining strips.

▶ Papering Casement Windows with Vinyl Wallpaper – Long Overlap

1. Paste and hang strip [3]. Trim strip at ceiling and baseboard.
2. Cut and remove piece [6] from strip [3], being careful to leave enough length on strip at top [4] and bottom [7] to reach window frame. Add 1 inch extra for trimming at window frame.
3. Beginning at top corner [2] make a 1-inch slit at 45° angle.
4. Make vertical cut [1] from edge of slit to edge of strip.
5. Smooth flap [5] against casement. Trim excess wallpaper at window frame.
6. Beginning at bottom corner [9], make a 1-inch slit at 45° angle.
7. Make vertical cut [8] from edge of slit to edge of strip.
8. Smooth flap against casement. Trim excess wallpaper at window frame.

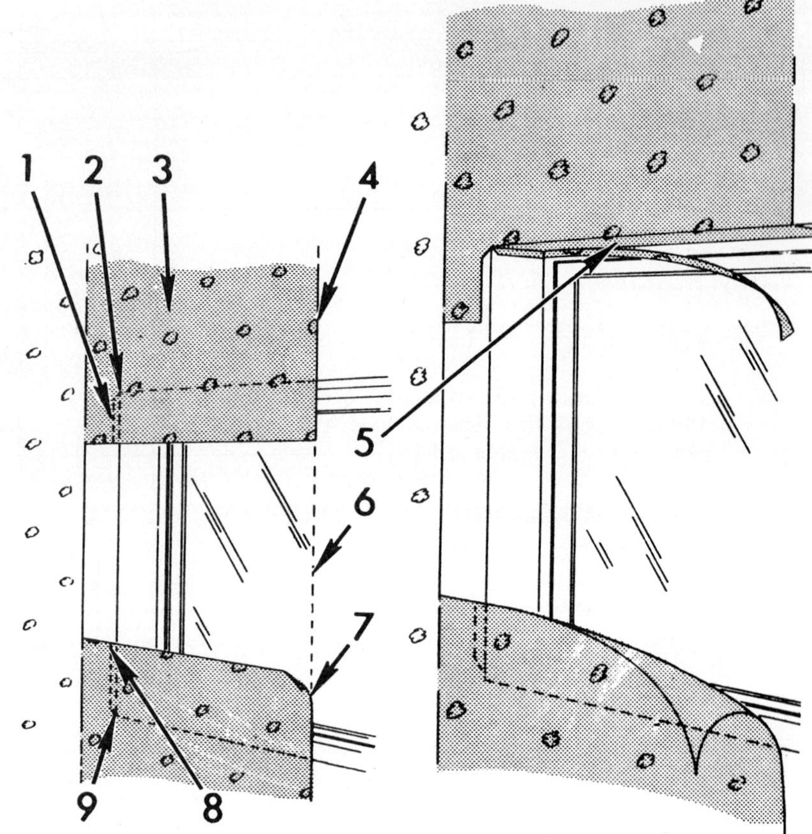

9. From a new piece of wallpaper, measure and cut a matching piece [2] of wallpaper. Width [3] of piece must be from edge of strip to window frame. Add 1 inch to width to allow for adjustment and trimming.
10. Paste, hang and trim piece [2].
11. Double cut overlap joints [1,4]. Go to Page 19 for making a double cut.
12. Match, cut and hang remaining strips.

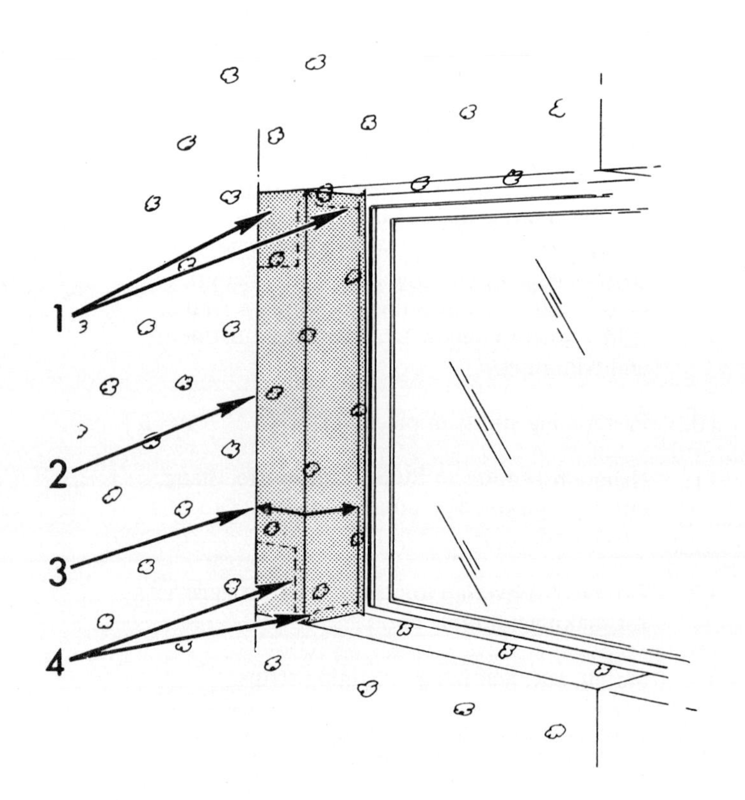

▶ Papering Casement Windows with Wallpaper Other than Vinyl

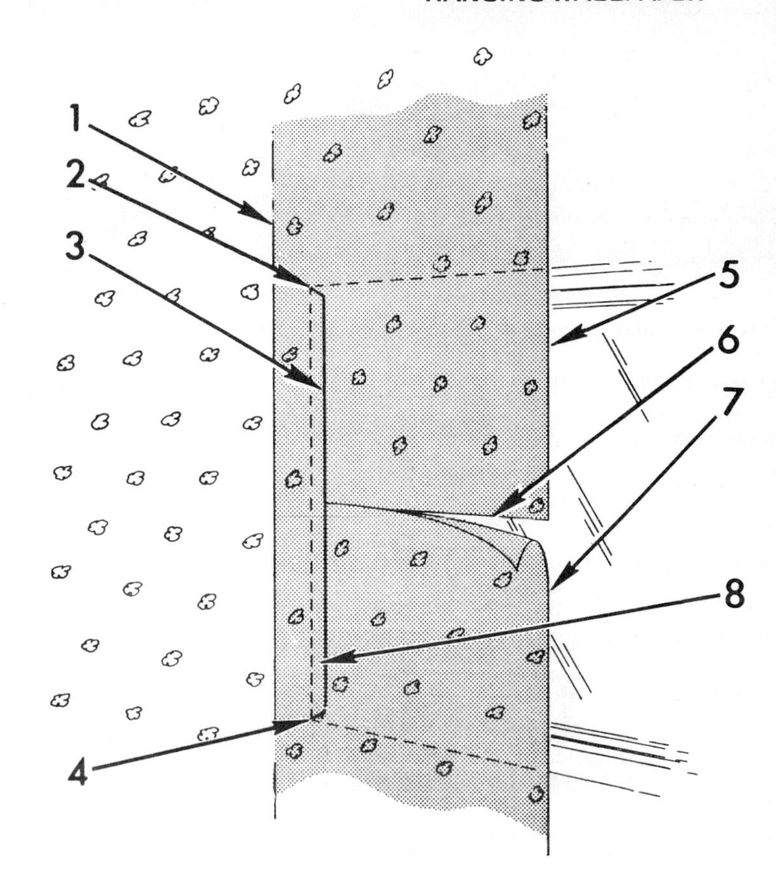

1. Paste and hang strip [1]. Trim strip at ceiling and baseboard.
2. Beginning at top corner [2] make 1-inch slit at 45° angle.
3. Beginning at bottom corner [4] make 1-inch slit at 45° angle.
4. Beginning 3/4-inch from edge of casement, make vertical cut [3] from bottom 45° slit to top 45° slit.
5. Make cut [6], being sure that there is enough paper above and below cut to reach window frame.
6. Smooth flap [8] against casement.
7. Smooth flap [5] against casement. Trim excess wallpaper at window frame.
8. Smooth flap [7] against casement. Trim excess wallpaper at window frame.

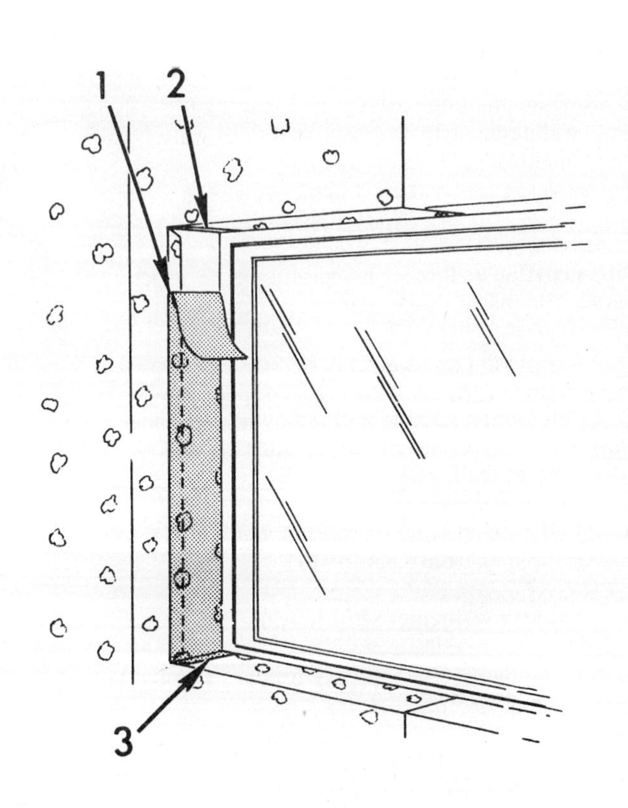

9. From a new piece of wallpaper, measure and cut a matching piece [1] of wallpaper. Matching piece [1] must be long enough to join or overlap top flap [2] and bottom flap [3].

Depending upon overlap of wallpaper with window, width of piece [1] may not be enough to reach window frame. If piece [1] does not reach window frame, cut and hang an additional matching piece to complete papering.

10. Paste, hang and trim piece [1].
11. Match, cut and hang remaining strips.

▶ Papering Around Electrical Switches and Outlets

It is convenient to remove cover plates from all switches and outlets on surfaces to be papered before beginning papering.

Cover plates are often covered with matching wallpaper to make them blend with background. Cut matching piece of wallpaper from scraps. Pattern must be matched to pattern around switch or outlet. Cut piece slightly larger than cover plate so that edges can be wrapped around edges of cover plate. Piece can be applied with wallpaper paste or white glue.

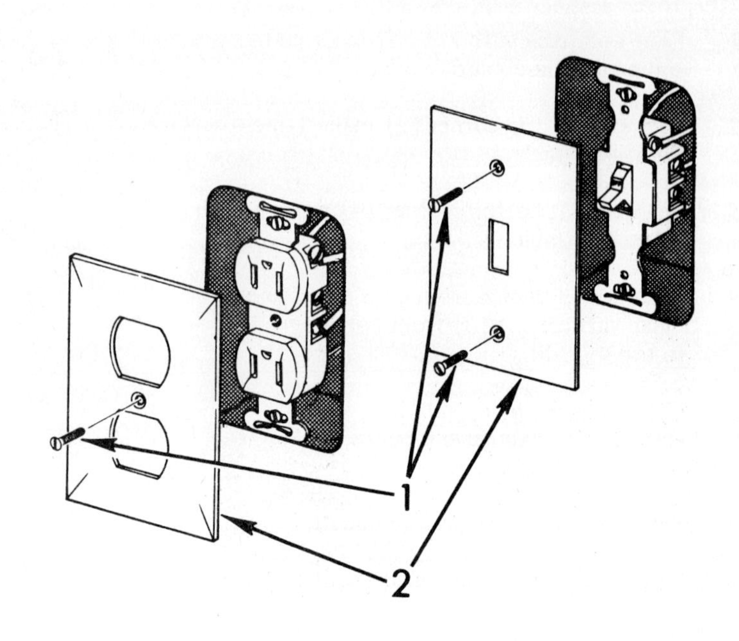

WARNING

Removal of cover plates will make it easier to touch "hot" wires. Therefore, if children are around, remove cover plates only at the time necessary for papering around the switch or outlet.

1. Remove cover plate [2] by removing screws [1].

Papering Around Electrical Switches and Outlets

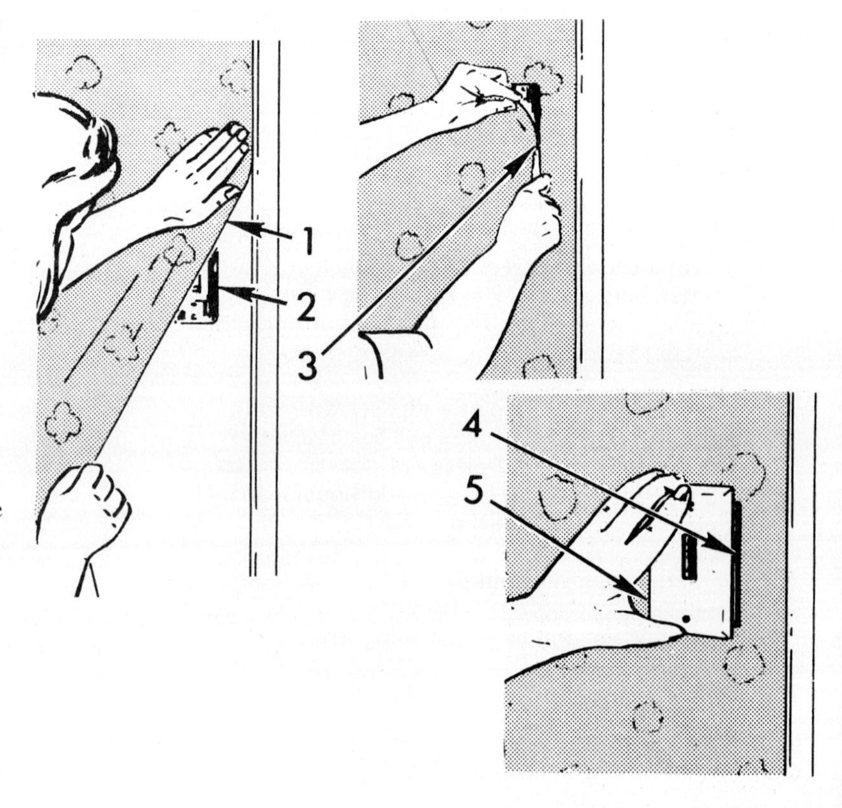

2. While aligning patterns and seam, hang wallpaper strip [1] over switch or outlet opening [2].

WARNING

When cutting wallpaper be careful not to touch wires. Electrical shock and injury may result.

CAUTION

Be careful when cutting and removing wallpaper from opening. Cover plate [4] must cover edges [5] of wallpaper.

3. Using sharp knife or scissors, cut and remove wallpaper from opening [3].

4. Smooth wallpaper strip to wall.

5. Using damp cloth, press edges [5] firmly around opening.

6. Place cover plate [4] at installed position. Install screws.

▶ Papering Around Light Fixtures

There are two methods for papering around light fixtures:

- Removing only the cover plate [3] before papering
- Removing the fixture [2] before papering.

If only the cover plate is removed, the wallpaper must be cut [1] to go around the fixture. This cut [1] will result in a seam which may be visible if it cannot be covered by the cover plate. However, it is by far the fastest way to paper around a fixture and is generally adequate.

The cut [1] may be visible because of its length and because of the shadows cast by direct lighting. If the cut would result in a bad appearance, it is best to remove the fixture. Although the procedure is time consuming, it results in the best appearance. Because the procedure is somewhat difficult, it is described here.

Although there are many types and styles of light fixtures, the method of installation is basically the same. Instructions apply to wall fixtures as well as ceiling fixture.

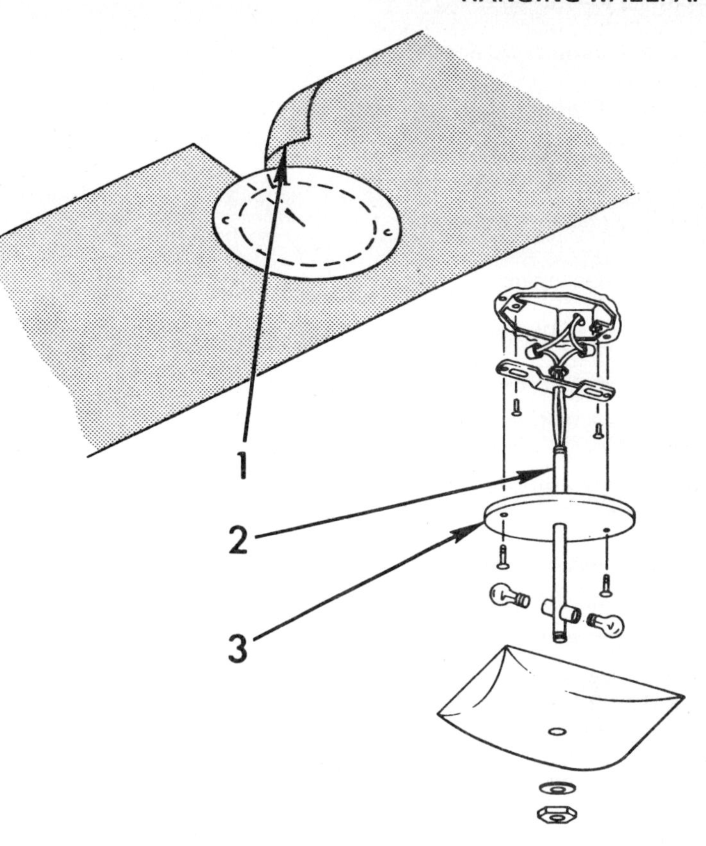

Papering Around Light Fixtures

WARNING

Do not remove light fixtures until electrical power is turned off. Power is turned off at fuse box or circuit breaker box.

1. While holding fixture cover [9], remove nut [10]. Remove cover.

2. Remove light bulbs [8].

3. Remove cover plate [6] by removing screws [7].

4. Remove mounting screws [5]. Pull fixture from opening [1].

When disconnecting wires [2], label each wire to aid in installation of fixture.

Solderless connectors [3] are removed by turning counterclockwise.

5. Label and disconnect wires [2,4] by removing solderless connectors [3]. Remove fixture. Push wires into opening [1].

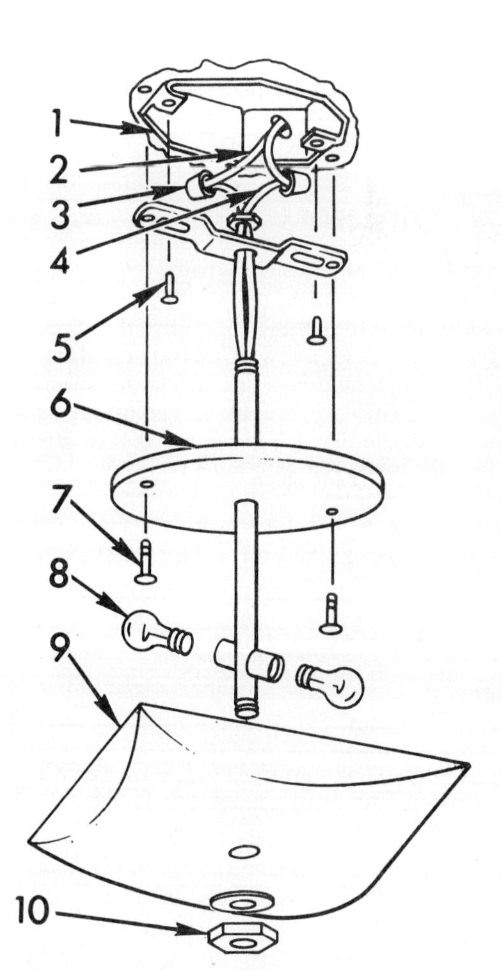

HANGING WALLPAPER

Papering Around Light Fixtures

6. While aligning pattern and seam, hang wallpaper strip [1] over opening [2].

7. Using sharp knife, cut and remove wallpaper from opening [2].

8. Smooth wallpaper strip to surface.

9. Using damp cloth, press edges firmly around opening.

Solderless connectors [4] are installed by placing over ends of wire and turning clockwise. Push solderless connector firmly onto wires while turning it.

10. Pull wires from opening [2]. Connect wires [3, 5] by installing solderless connectors [4]. Remove labels.

11. Place fixture at installed position. Install mounting screws [6].

12. Place cover plate [7] at installed position. Install screws [8].

13. Install light bulbs [9].

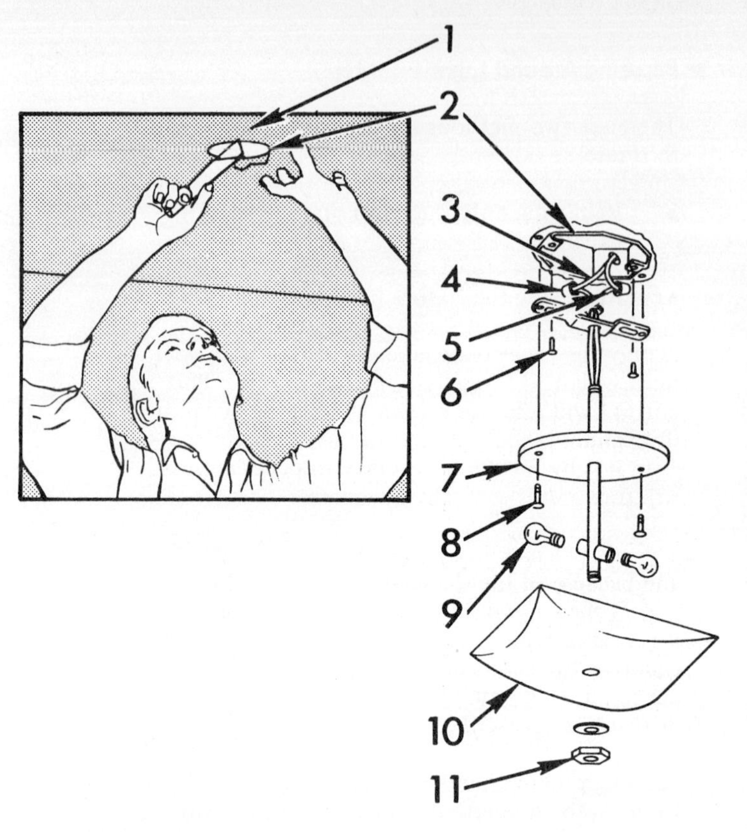

14. Place fixture cover [10] at installed position. Install nut [11].

HANGING MURALS

Murals are printed on paper, vinyl or foil materials. The special handling requirements for these materials apply to murals as well as to wallpapers. See Page 2 for the handling requirements for different wallpaper materials.

In addition to the requirements listed in these pages, murals have a few of their own. Murals are used like large paintings rather than just wallcoverings. They will receive special notice by others and, therefore, will require special attention to install. Particular care must be given:

- to preparing the wall surface for the mural and
- to determining the proper location for the mural on the wall

It is recommended that the wall surface be sanded to remove any texture and roughness. It is also recommended that backing paper (blank stock) first be applied to the wall to provide a smooth surface and better bonding for the mural. Backing paper is an unpatterned paper which is available at wallpaper dealers. Ask the dealer for advice regarding use of backing paper.

Backing paper must be applied with the same adhesive which is used for the mural. Vinyl adhesives are strongly recommended to ensure good bonding and to prevent mildew which will damage the mural.

A difficult part of hanging a mural is to determine the proper location for it. Individual preferences, type of figure, furniture arrangement and use of and layout of the room all play a part in selecting the proper wall for the mural and the proper location on the wall for the mural.

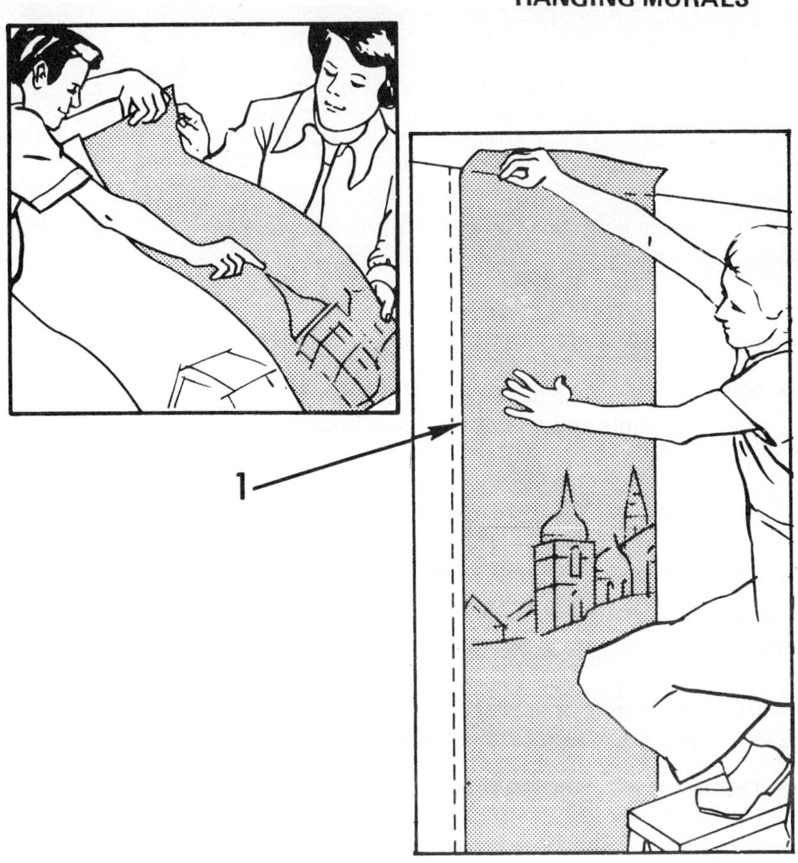

In general, a mural is located so that its centerline is on the centerline of the wall.

The up and down location of the mural between the ceiling and the floor is best determined by trial and error. Find the panel [1] which contains the highest part of the figure.

While holding this panel [1] against the wall, move it up and down until it looks right.

Murals are seldom large enough to fill an entire wall. Therefore, special background (companion) wallpapers are available for completing the wall or, if desired, all remaining walls. Your wallpaper dealer can help you select proper background wallpaper.

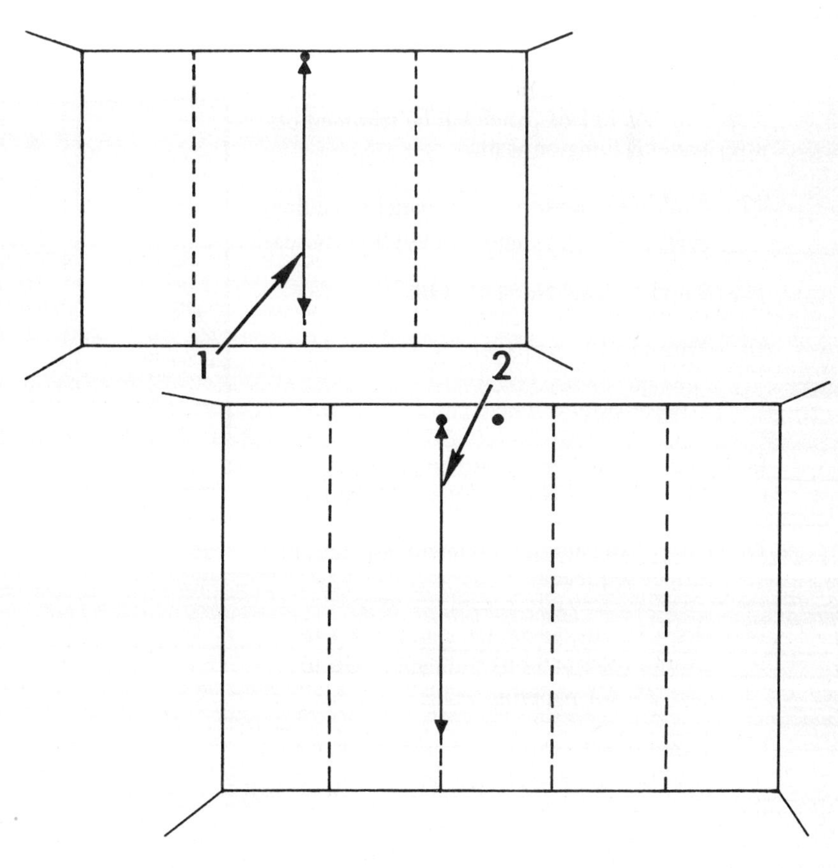

Tools and supplies which are required to hang murals are the same as those which are required to hang wallpaper. In addition, you may need backing paper (blank stock).

Use hairline joint seam, Page 18, for backing paper. Procedures for hanging wallpaper apply to backing paper. Go to Page 25 for papering walls.

1. Determine centerline of wall and mark lightly with pencil.

2. If mural has even number of panels, mark plumb line [1] on centerline. Go to Page 11 for making a plumb line.

3. If mural has odd number of panels, mark plumb line [2] 1/2 width of a panel to the left of the centerline. Go to Page 11 for making a plumb line.

4. While holding panel containing highest part of figure against wall, move panel up and down until preferred height of figure is determined.

5. Make light mark on panel to show where top of panel meets ceiling.

6. Make light mark on panel to show where bottom of panel meets baseboard.

7. Add 2 inches to top of panel to allow for adjusting and trimming. Using straightedge and knife, cut top of panel.

8. Add 2 inches to bottom of panel to allow for adjusting and trimming. Using straight-edge and knife, cut bottom of panel.

9. Check that panel is cut correctly, by holding in position against wall.

10. Match remaining panels to panel. Cut tops and bottoms of remaining panels.

Procedures for hanging wallpaper apply to hanging murals. Before doing next step, read:

- Making Seams – Butt Joint Seams, Page 18.
- Trimming Wallpaper, Page 20.
- Mixing paste, Page 21.
- Papering Walls, Page 25.

Panels should be hung in sequence as illustrated.

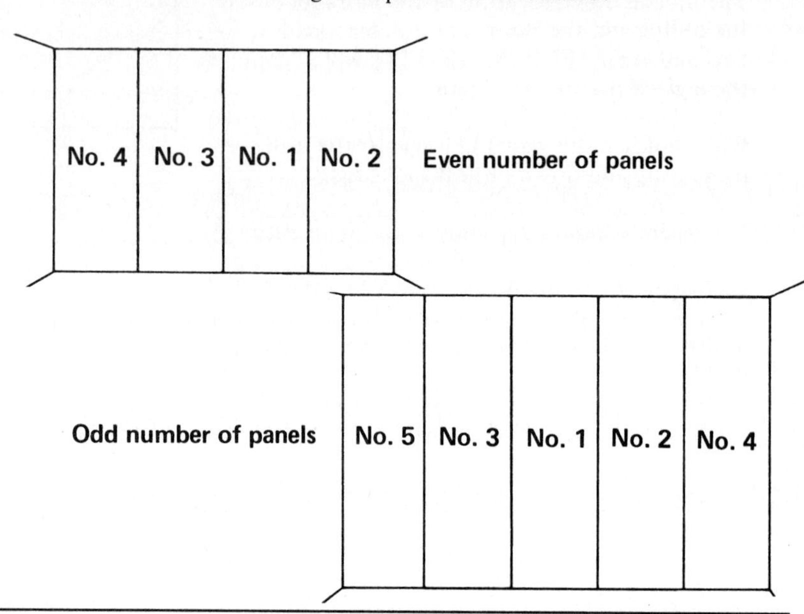

11. Apply paste to panel No. 1.

Top 2 inches [1] of panel will be trimmed off after panel is hung on wall.

12. While aligning top [2] of panel to ceiling, align left edge of panel to plumb line [3].

13. Smooth panel No. 1 to wall.

14. Trim top [2] of panel No. 1.

15. Trim bottom of panel No. 1.

Left edge of panel No. 2 will be placed against right edge of panel No. 1, using butt joint seam.

16. Paste, hang and trim remaining panels in proper sequence.

17. Measure, cut, paste, hang and trim background (companion) wallpaper. Go to Page 25, for papering walls.

If only mural wall is to be papered, background paper will be trimmed flush at each corner of the wall.

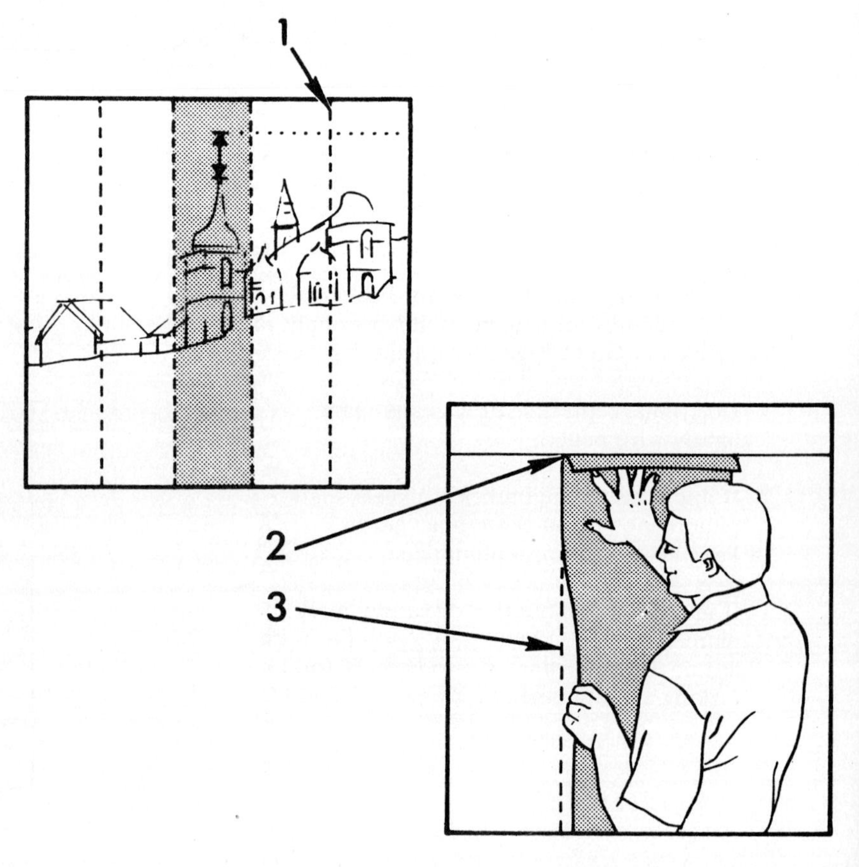

Instructions in this section tell how to store paste and cleanup tools and work area.

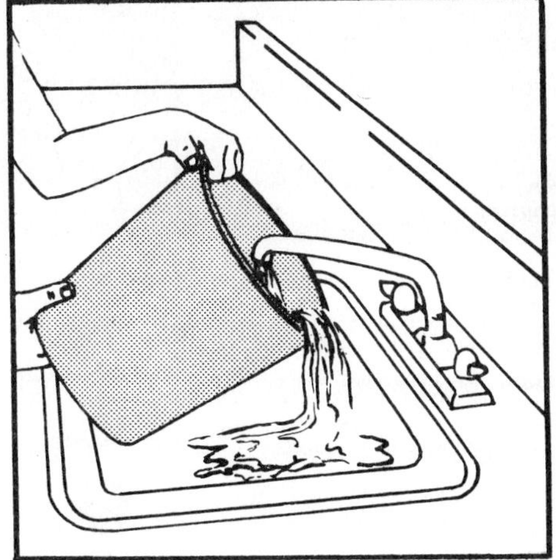

▶ Storing Paste

If storing premixed paste, pour unused paste back into its purchase container. Seal the container tightly.

If storing dry mix paste, place unused dry mix into an airtight container. Seal the container tightly. Label the container and store it in a dry place.

Paste made from dry mix is organic and biodegradeable. It is not harmful to sewage systems. Pour unused dry-mix paste down drain.

▶ Cleaning Tools

1. Using warm water, rinse bucket, pail, or roller tray until water runs clear.

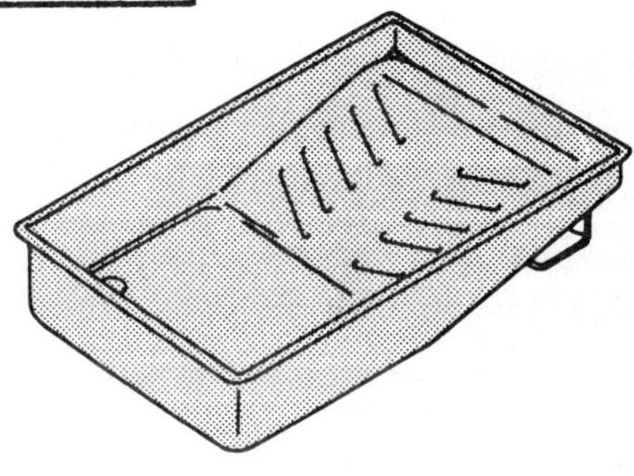

2. Using warm, soapy water, wash bucket, pail, or roller tray. Rinse in warm water until water runs clear. Allow to dry completely before storing.

Cleaning Tools

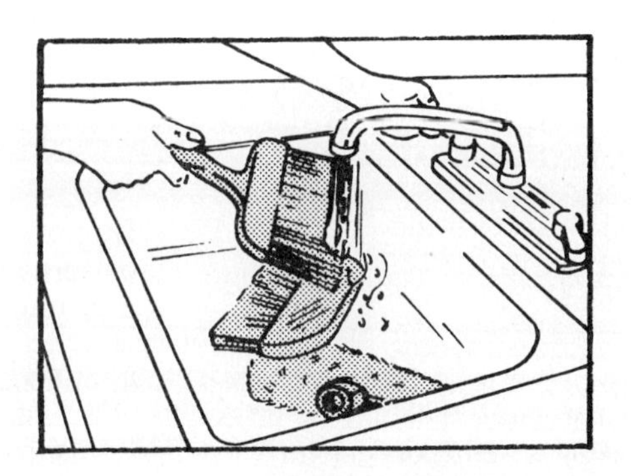

3. Using warm, soapy water, wash brushes and rollers to remove any paste buildup. Rinse in warm water until water runs clear. Allow to dry thoroughly before storing.

4. Using sponge and warm water, clean remaining tools. Wipe tools dry before storing.

▶ Cleaning Work Area

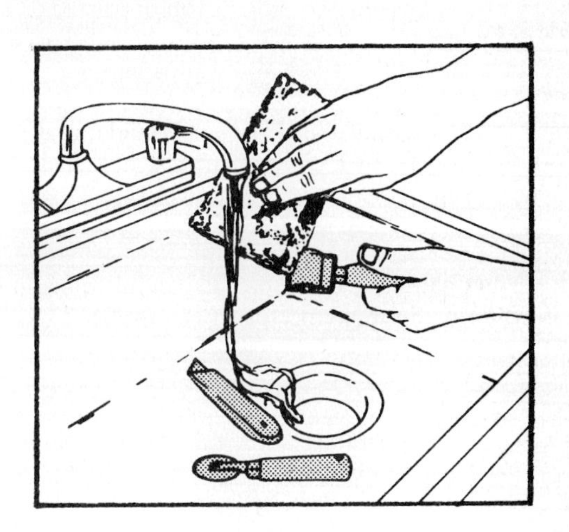

1. Remove paper covering from pasting table. Using sponge and clean water, clean table.

2. Wipe off any paste from drop cloths. Fold and store drop cloths in dry area.

3. Replace any wall fixtures removed during preparation for wallpapering.

▶ Repairing Loose Edges, Peeling, and Small Tears

The following tools and supplies are required:

Seam roller [1]
Brush [2]
Sponge or clean rag
White glue or wallpaper paste

CAUTION

When handling wallpaper be careful to avoid sharp folding. Permanent creases may result.

1. While holding paper away from surface, apply glue or paste to surface and wallpaper.

CAUTION

Do not use seam roller on flocked or raised-patterned wallpaper. Damage to flock or pattern may result.

2. Place wallpaper against surface. Using seam roller or fingers, press wallpaper tightly into place.

3. Using damp sponge, remove excess glue or paste from wallpaper. Be careful not to soak wallpaper.

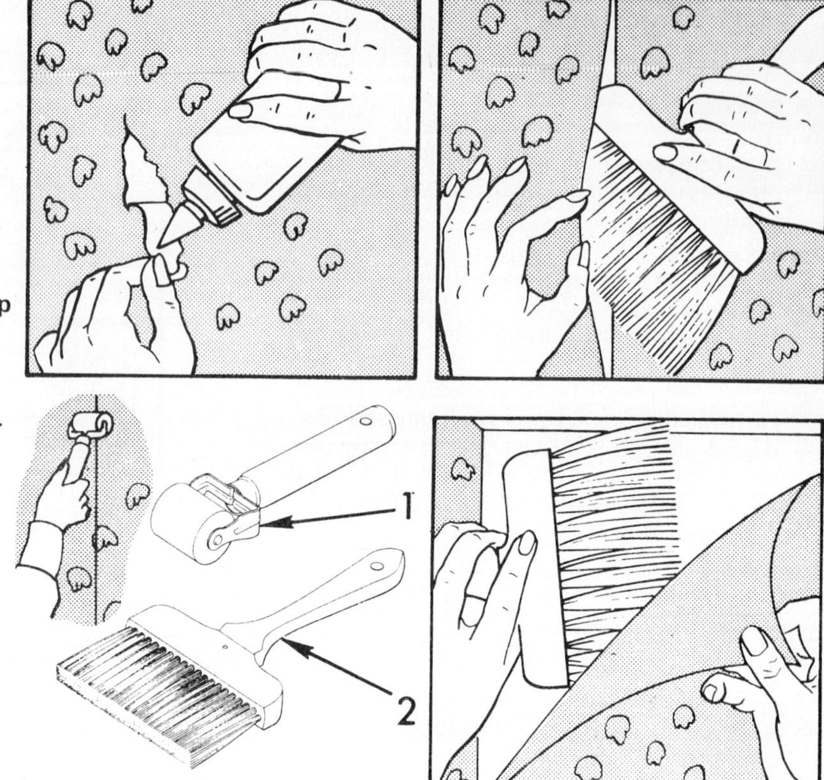

▶ Repairing Blisters and Bubbles

The following tools and supplies are required:

Sharp knife [1] or razor blade
Seam roller [2]
Sponge or clean rag
White glue

When cutting wallpaper, cut [3] should be made along pattern if possible to help conceal cut. Cut [3] should be made at top of blister or bubble so that glue will run down into space between wall surface and wallpaper.

1. Using sharp knife or razor blade, make an X-, T-, or L-shaped cut [3] in bubble.

CAUTION

When handling wallpaper be careful to avoid sharp folding. Permanent creases can result.

2. Carefully lift cut edges [4].

3. While holding cut edges [4] away from surface, apply glue to surface and wallpaper.

CAUTION

Do not use seam roller on flocked or raised-patterned wallpaper. Damage to flock or pattern may result.

4. Place wallpaper against surface. Using seam roller or fingers, press wallpaper tightly into place.

5. Using damp sponge, remove excess glue from wallpaper. Be careful not to soak wallpaper.

1

2

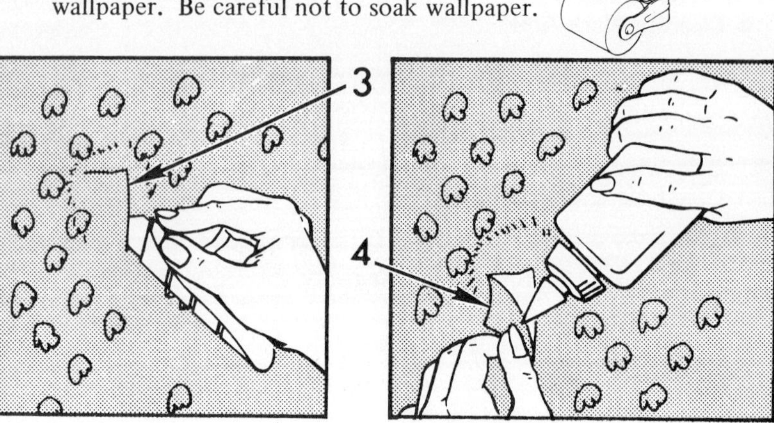

▶ Repairing Large Tears and Damaged Areas

The following tools and supplies are required:

Metal straightedge [1]
Sharp knife [2] or razor blade
Putty knife [3]
Brush [4]
Seam roller [5]
Piece of matching wallpaper
Masking tape
Sponge or clean rag
Wallpaper paste
or white glue

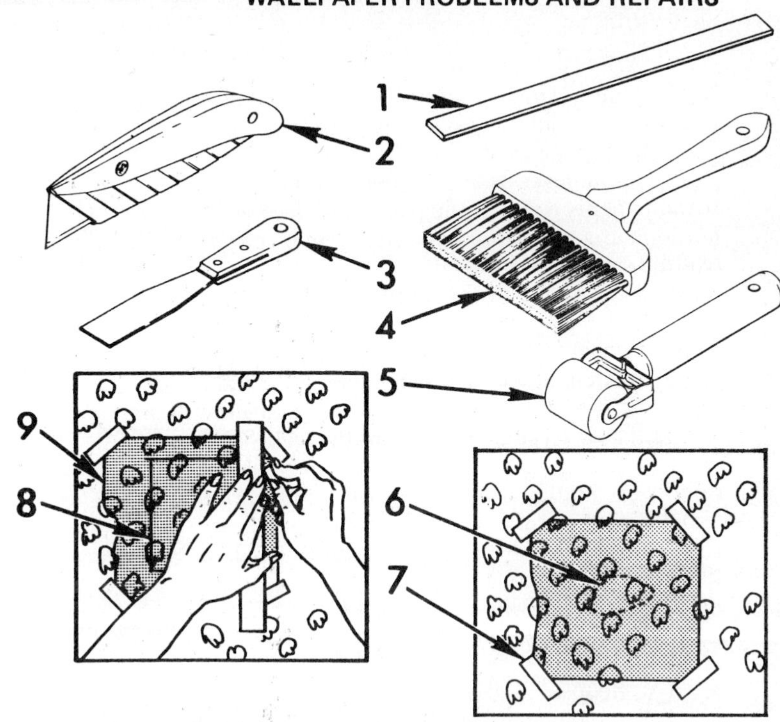

Make sure that damaged area [6] will be in center of new piece of wallpaper when patterns are aligned.

1. Cut a piece of matching wallpaper slightly larger than damaged area [6].

2. While holding new piece at aligned position, lightly apply masking tape [7] at four corners.

When cutting wallpaper, be sure to cut through wallpaper on wall.

If wallpaper pattern has straight lines, cuts should be made on straight lines so cuts will not be noticeable.

Make sure that damaged area [6] will be inside of cut area.

3. Using straightedge and sharp knife, cut rectangular patch [8] from new piece.

4. Remove patch [8]. Remove cut piece [9] by carefully removing tape.

Repairing Large Tears and Damaged Areas

CAUTION

Do not soak surrounding wallpaper when wetting damaged area.

3. Using wet sponge, thoroughly soak damaged area.

CAUTION

When using putty knife, be careful not to damage wall surface or surrounding wallpaper.

4. Using putty knife, scrape damaged wallpaper [1] from wall.

5. Using clean water, thoroughly wash exposed wall surface [2] to remove all old paste. Allow surface to dry completely.

6. Check surface of wall for damage.

7. Apply glue or paste to patch [3].

CAUTION

Do not use seam roller on flocked or raised-patterned wallpaper. Damage to flock or pattern may result.

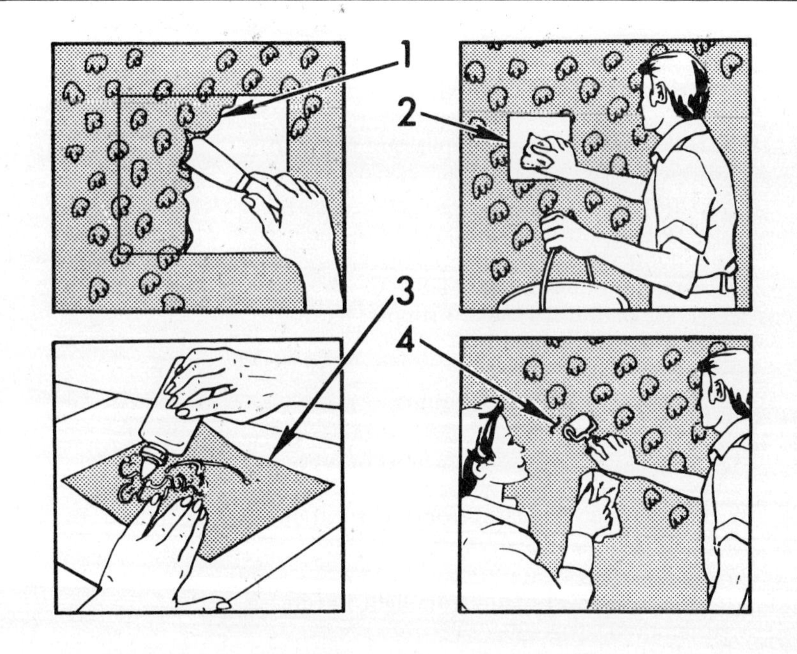

8. Place patch [3] against clean wall surface [2]. Using seam roller or fingers, press patch tightly into place [4].

9. Using damp sponge, remove excess glue or paste from wallpaper. Be careful not to soak wallpaper.

Before painting or re-papering a wall or ceiling, you should remove any old wallpaper from it first.

Paint should never be applied over wallpaper. If wallpaper is painted, the painted surface will not look as good as it could. Also, the paint will seal the surface of the wallpaper and make it very difficult to remove in the future.

Applying new wallpaper over old wallpaper is not recommended either. The glue holding the old wallpaper to the wall may become wet and allow the wallpaper to come loose from the wall.

In no case should vinyl wallpaper be applied over old wallpaper. If the old wallpaper is coated or vinyl, the glue between them will not dry well. If the old wallpaper is not vinyl, the glue under the old wallpaper will become wet and may start to mildew.

Many wallpapers used in recent years are strippable. They are designed to be easily removable from walls and ceilings. If your wallpaper is strippable, go to next section (below) for removal.

Non-strippable wallpapers are not designed for easy removal. They are very difficult to remove unless special equipment or materials are used. The job of removing non-strippable wallpaper is made much easier by use of special steamers or by use of chemical wallpaper removers.

Steamers can, if not used carefully, soften the paper surface on wallboard. The wallboard can then be damaged by sharp putty knives or other tools used to scrape the wallpaper from the wallboard. Therefore, chemical removers are generally recommended for the novice.

If your wallpaper is non-strippable, go to Page 45 for removal by using a steamer. Go to Page 46 for removal by using chemical removers.

▶ Removing Strippable Wallpaper

The following tools and supplies are required:

- Drop cloths
- Container
- Sponges or rags
- Trisodium phosphate (T.S.P.).
  Available at builder's supply, hardware store, or paint store

1. Remove or protect furniture.
2. Cover floors and carpets in work area.
3. Remove cover plates from electrical outlets and switches as required. Page 36.
4. Remove cover plates from light fixtures as required. Page 37.

Wallpaper should be removed one strip at a time.

5. Beginning at one corner of strip, carefully pull strip from wall. Repeat for all strips.

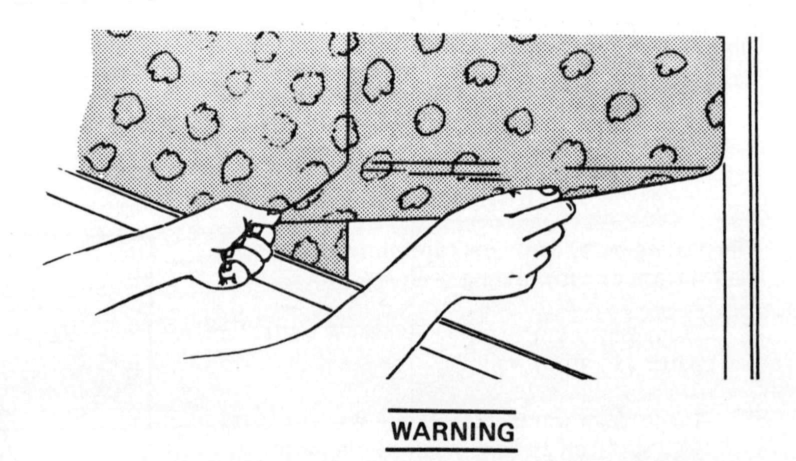

WARNING

Be careful to keep water from entering electrical openings. Electrical shock and injury could result.

6. Following manufacturer's instructions, mix a solution of trisodium phosphate (T.S.P.) and water. Thoroughly wash surface. Rinse with clear water. Allow to dry completely.

▶ Removing Non-Strippable Wallpaper with a Steamer

The steamer is used to soften the glue between the wallpaper and the wall. Ask the wallpaper dealer for instructions for using the steamer.

If wallpaper is vinyl coated, it may be necessary to cut many small holes into wallpaper so that steam can reach the glue. A special tool is available for doing this. Ask your wallpaper dealer for this tool and instructions for using it.

The following tools and supplies are required:

Steamer [1]. Available from wallpaper store
Putty knife [2]. Wide, slanted blade type is recommended.
A puncturing roller [3] is recommended for making holes in nonporous wallpapers. The holes allow the steam or remover solution to penetrate under the wallpaper.
Drop cloths
Container
Sponges or rags
Trisodium phosphate (T.S.P.). Available at builder's supply, hardware store or paint store

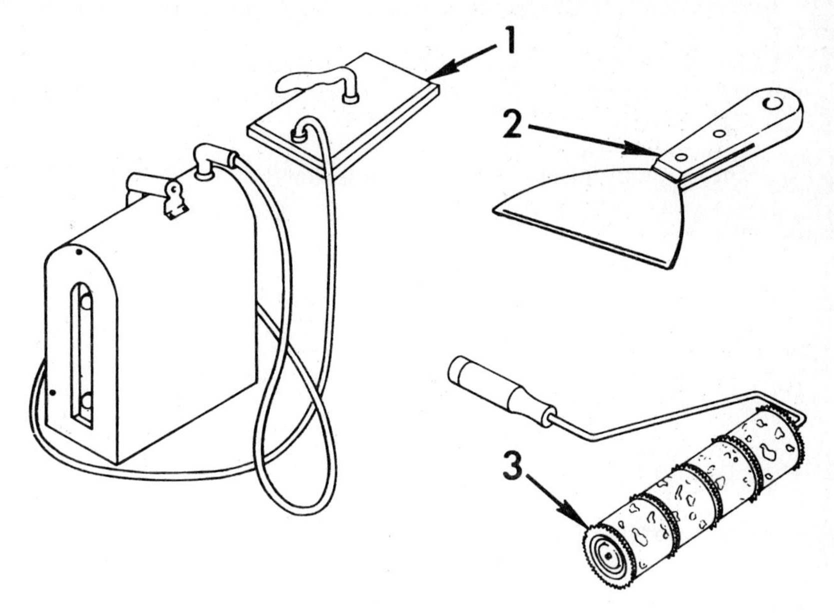

1. Remove or protect furniture.
2. Cover floors and carpets in work area.
3. Remove cover plates from electrical outlets and switches as required. Page 36.
4. Remove cover plates from light fixtures as required. Page 37.

Removing Non-Strippable Wallpaper with a Steamer

Wallpaper should be removed one strip at a time.

Begin at top of strip and work down to bottom of strip.

CAUTION

When using putty knife, be careful not to damage wall surface.

5. While applying steam to strip of wallpaper, carefully remove wallpaper from wall with putty knife. Repeat for each strip.

WARNING

Be careful to keep solution from entering electrical openings. Electrical shock and injury could result.

6. Following manufacturer's instructions, mix a solution of trisodium phosphate (T.S.P.) and water. Thoroughly wash surface. Rinse with clear water. Allow to dry completely.

REMOVING WALLPAPER

▶ Removing Non-Strippable Wallpaper with Chemical Removers

Chemical removers dissolve the glue between the wallpaper and the wall. Ask the wallpaper dealer for instructions for applying chemical remover.

The following tools and supplies are required:

Putty knife [1]. Wide, slanted blade type is recommended

Puncturing roller [2]. A puncturing roller is recommended for making holes in wallpapers. The holes allow the remover solution to penetrate under the wallpaper. Available at wallpaper stores.

Drop cloths — Sponges or rags

Container — Chemical remover

Trisodium phosphate (T.S.P.). Available at builder's supply, hardware store or paint store

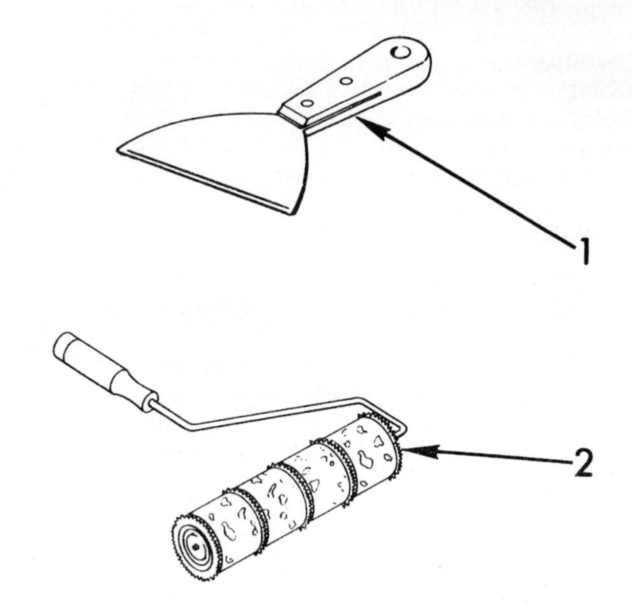

1. Remove or protect furniture.

2. Cover floors and carpets in work area.

3. Remove cover plates from electrical outlets and switches as required. Page 36.

4. Remove cover plates from light fixtures as required. Page 37.

WARNING

Be careful to keep chemical remover from entering electrical openings. Electrical shock and injury could result.

5. Following manufacturer's instructions, apply chemical remover to wallpaper.

CAUTION

When using putty knife, be careful not to damage wall surface.

6. Using putty knife, carefully remove wallpaper from wall.

WARNING

Be careful to keep solution from entering electrical openings. Electrical shock and injury could result.

7. Following manufacturer's instructions, mix a solution of trisodium phosphate (T.S.P.) and water. Thoroughly wash surface. Rinse with clear water. Allow to dry completely.

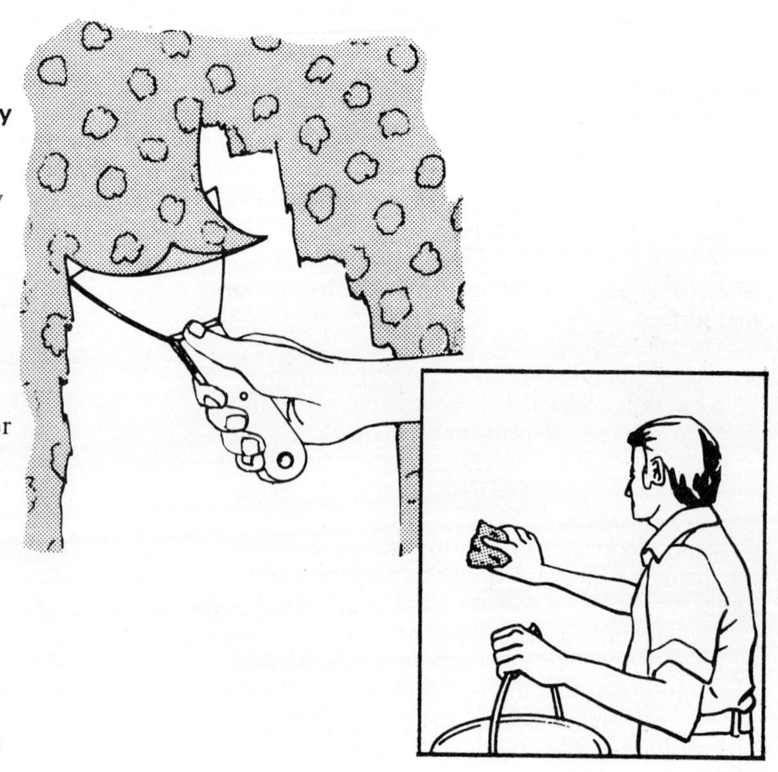

▶ Protecting Wallpaper

For easier cleaning and longer service life, wallpaper may be given a protective coating. These coatings are especially useful for protecting paper-surfaced wallpaper in traffic areas such as hallways or around light switches.

Protective coatings are available which dry to a clear washable surface. They do not discolor the wallpaper.

Ask a wallpaper dealer to recommend the protective coating to be used with a particular type of wallpaper.

CAUTION

Do not apply protective coating to newly hung wallpaper until paste is completely dry. Mildew can occur on wall surface.

1. Following manufacturer's instructions, apply protective coating to wallpaper.

When coating is dry, wallpaper surface is completely washable.

▶ Cleaning with Commercial Dough

Commercial cleaning dough is available to remove dirt and grime from washable and non-washable wallpaper. Cleaning dough is not used for spot cleaning. It is used for cleaning entire walls or rooms.

CAUTION

Always test commercial cleaning dough on a hidden part of the wallpaper before applying it to soiled area. Cleaning dough may affect the color or texture of wallpaper. Do not use cleaning dough if wallpaper is affected.

When using cleaning dough, begin at the top of the surface and work down, using long, light strokes. Make all strokes in same direction to avoid streaking.

As cleaning dough becomes soiled, it must be kneaded until clean.

1. Using cleaning dough, clean surface.
2. After cleaning, wipe surface with clean rag to remove any dough remaining on wallpaper.

▶ Removing Stains on Washable Wallpaper

Dirt, grease, and stains must be removed before they penetrate into the wallpaper.

CAUTION

If there is doubt about washability of wallpaper, test on a hidden part of the wallpaper.

1. Thoroughly wash soiled area with mild soap and cold water solution. Rinse with clear, cold water. Wipe dry with clean, absorbent cloth.

If stain remains, go to last paragraph to use commercial spot remover to remove stain.

▶ Removing Stains on Non-Washable Wallpaper

Dirt, grease, and stains must be removed before they penetrate into wallpaper.

CAUTION

When cleaning non-washable wallpaper, be sure to moisten soiled area only.

1. Using sponge moistened with mild soap and cold water solution, carefully blot soiled area.
2. Using sponge moistened with clear, cold water, carefully blot soiled area to remove soap. Blot dry with clean, absorbent cloth.

If stain remains, go to next paragraph to use commercial spot remover to remove stain.

▶ Removing Stains with Commercial Spot Removers

Commercial spot removers are available to remove stains from different types of wallpaper.

Ask a wallpaper dealer to recommend a spot remover for the type of wallpaper to be cleaned.

CAUTION

Always test spot remover on a hidden part of the wallpaper before applying it to soiled area. Spot remover may affect the color or texture of wallpaper. Do not use spot remover if wallpaper is affected.

Following manufacturer's instructions, clean stained area with commercial spot remover.

Creative decorating with wallpaper and paint

Fern-patterned wallpaper provides the background for this cozy living room, with accents supplied by drapes and pillows in the matching fabric. Like ''Petite'' (shown here) from Stauffer's Miniatures collection, many wallcoverings have fabrics to match — adding a new dimension to decorating possibilities. Design: Decorage. Photography: Ted DeJony. *Photo courtesy of Stauffer Chemical Company*. (left)

A bold checkerboard floor provides the focus for this unusual room. The border to the black-and-white area is painted a bright canary-yellow to match the fabric coverings on the chairs and the pillow-backed benches. The white-painted brick wall highlights the white wicker furniture. The natural bamboo shades and the generous assortment of plants subtly mute the starkness of the geometrical theme. *Photo courtesy of Pittsburgh Paints*. (below)

Wallpaper makes a living room or den unique.

The handsome and timeless look of wood and leather is at home against the brown and beige ''Pinehurst'' pattern, a prepasted vinyl from the Len-Fast collection. Trim is painted a coordinating brown; the off-white lampshade and sheers lighten the decor. *Photo courtesy of Lennon Wallpaper Company, Joliet, Illinois.* (left)

In the tradition of Lord Paisley, who brought back the first ''paisley'' pattern from the Near East, Oleg Cassini has designed the ''Cassini Paisley'' for Birge's Oleg Cassini collection. Here it adds a splash of pattern to a den that features carpet-covered platform seating. The raw pine trim adds country comfort to a room that's meant for restful reading. Design: Dorothy Wyeth Dobbins. Photography: Photographic House. *Photo courtesy of Birge Wallcoverings, Inc.* (below left)

The futuristic motif of this living room is enhanced by the silver-toned ''Standing Up'' pattern that blends beautifully with chrome accessories. From the Home Fashions collection. *Photo courtesy of Columbus Coated Fabrics, Chemical Division, Borden Inc.* (below right)

Distinctiveness = Wallpaper + Sofa

Guaranteed to add pizzazz to any wall is ''Interlude,'' a foil covering from the Wall-Tex Metallics collection. The textured fabric of the sofa picks up the color of one of the wall stripes. *Photo courtesy of Columbus Coated Fabrics, Chemical Division, Borden Inc.* (above left)

Quiet and casual is the mood of this pattern of stylized branches. Rough wood, pussy willows, and corduroy complete the look of graceful comfort. *Photo courtesy of Birge Wallcoverings, Inc.* (above right)

Oriental harmony prevails in this restful setting with a backdrop of ''Batik,'' designed by Diane von Furstenberg for the Wall-Tex collection. *Photo courtesy of Columbus Coated Fabrics, Chemical Division, Borden Inc.* (right)

Focus on the fireplace

Unquestionably traditional, this fireside setting is ideal for the muted tints of Lennon's elegant ''Georgetown'' pattern from the Colonial Sampler collection. The matching fabric is shown to beautiful advantage in the full-length drapes and valance. *Photo courtesy of Lennon Wallpaper Company, Joliet, Illinois.* (right)

The same ingredients — wood fireplace and leather easy chair — combine with the warm red of ''Nassau Stripe'' to create a very different mood. The flocked pattern is from Birge's Colonial collection. *Photo courtesy of Birge Wallcoverings, Inc.* (below)

Birge's ''Jeremiah Stripe,'' available in four different color combinations, dresses up the plywood chaise in this living room. The vertical stripe is ''railroaded'' around the base of the hearth and platform that forms a ledge beneath the wood-screened windows. From the Oleg Cassini collection. *Photo courtesy of Birge Wallcoverings, Inc.* (above)

The perfect place to relax with a hot drink is this updated Victorian corner by the hearth. Painted mantel and moldings pick up the rich wine-red of Borden's ''Orient Express'' pattern from the Passport collection. *Photo courtesy of Columbus Coated Fabrics, Chemical Division, Borden Inc.* (left)

Wallpaper complements — or creates — a variety of styles in dining areas.

Antiques dress up with elegant wallcoverings from Stauffer's Miniatures collection. The natural pine window has a shade to match the "Minette" and "Minette Stripe" patterns. Design: Decorage. Photography: Ted DeJony. *Photo courtesy of Stauffer Chemical Company*. (facing page, top left)

And contemporary furniture goes country with Stauffer's coordinating "Provence/Provincial" patterns, also from the Miniatures collection. *Photo courtesy of Stauffer Chemical Company*. (facing page, bottom left)

The outdoor setting suggested by the wrought-iron table and chairs is enhanced by the live plants and the tree-patterned foil wallpaper ("Half Moon Bay" from the Wall-Tex Metallics collection). *Photo courtesy of Columbus Coated Fabrics, Chemical Division, Borden Inc*. (facing page, top right)

Formality is the statement of this dining arrangement — echoed by the flocked wallpaper and the plushly upholstered sofa. *Photo courtesy of Birge Wallcoverings, Inc*. (left)

Wicker furniture and the waving palms of Lennon's vinyl "Caribbia" pattern are the basic ingredients of this tropical mood. Accessories — bamboo shade, macrame hanging, woven mat, and cork-like floor tiles — accent the decor. From the Len-Fast collection. *Photo courtesy of Lennon Wallpaper Company, Joliet, Illinois*. (below left)

The striking juxtaposition of formal with casual is achieved by combining Borden's "Harbor Town" and "Calibogue Cay" patterns (both from the Passport collection). *Photo courtesy of Columbus Coated Fabrics, Chemical Division, Borden Inc*. (below right)

Cheery vinyl wallcoverings set apart a corner of the kitchen.

The comfortable but uncluttered look of this dining alcove starts with ''Madras,'' a plaid from Stauffer's Inspirations II collection. The wallpaper's dark and light hues are picked up by the painted surfaces. *Photo courtesy of Stauffer Chemical Company.* (above left)

Brightening the breakfast nook are Lennon's coordinated ''Country Kitchen'' and ''Country Stripe'' from the Len-Tex collection. ''Country Kitchen'' hues are picked up by the green-painted chairs and red-checked tablecloth and window-seat cover. *Photo courtesy of Lennon Wallpaper Company, Joliet, Illinois.* (above right)

Where home and hobby meet, ''Patches'' from Stauffer's Main Street II collection ties it all together. Hanging lamps, along with painted walls and moldings, accent the wallpaper colors. *Photo courtesy of Stauffer Chemical Company.* (right)

Bathrooms — plain and fancy

Serenity is the keynote here. Pulling it all together are the "Chamois" and "Fern" wallcoverings from Stauffer's Inspirations II collection. Potted ferns add the finishing touch. *Photo courtesy of Stauffer Chemical Company*. (left)

Providing the background for this Oriental fantasy is Borden's "Ikebana" from the Passport collection. *Photo courtesy of Columbus Coated Fabrics, Chemical Division, Borden Inc*. (below)

Large or small, bedrooms are made for wallpaper!

The quiet harmony of this bed-sitting-room starts with the walls, papered in Stauffer's coordinating ''Polly'' and ''Pollyanna'' patterns from the Miniatures collection. Design: Peter Bradley and Jack McCurdy. Carpet: Bigelow. Bed shawl: Park Smith. Lamps: Stiffel. *Photo courtesy of Stauffer Chemical Company.* (above)

''English Chintz,'' a pretty mini-floral design from Birge's new Colonial collection (vol. 59), turns a cramped and dreary attic into a bright bower, great for family or company. Windows and window seat are dressed in the matching fabric; trim and built-in drawers are painted a contrasting lilac. Design: Dorothy Wyeth Dobbins. Photography: Photographic House, Inc. *Photo courtesy of Birge Wallcoverings, Inc.* (left)

Paper doesn't have to end with the walls. Lennon's ''Quilting Bee'' (from the Colonial Sampler collection) covers the ceiling as well as the walls, with matching fabric used for drapes and bedspread. *Photo courtesy of Lennon Wallpaper Company, Joliet, Illinois.* (above)

Demonstrating that a lot isn't always too much is this cozy bedroom, decorated entirely with paper and fabrics in the ''Provincetown'' pattern from Lennon's Colonial Sampler collection. *Photo courtesy of Lennon Wallpaper Company, Joliet, Illinois.* (right)

Mirroring the many moods of childhood...

For the princess of any age, Birge's flowered foil adds sparkle and charm to a thoroughly feminine boudoir. *Photo courtesy of Birge Wallcoverings, Inc.* (right)

A boy's simple furnishings are the perfect counterpoint to Lennon's lively ''Red Hot'' pattern from the Len-Tex collection. *Photo courtesy of Lennon Wallpaper Company, Joliet, Illinois.* (below)

Happy playtime memories start with easy-care ''Applique'' from Stauffer's Main Street II collection. *Photo courtesy of Stauffer Chemical Company.* (facing page, top)

For a girl who's growing, almost anything goes with versatile ''Calico Daisy'' from Stauffer's Main Street II collection. *Photo courtesy of Stauffer Chemical Company.* (facing page, bottom)

AMERICAN
MUSIC HALL
COMING ATTRACTIONS
25-50¢
Vaudeville
HELLO
SONGS

JAW'Y

Corners come alive with wallpaper.

An upholstered platform strewn with pillows makes this foyer a place to linger. Lennon's airy ''Lexington'' pattern (from the Colonial Sampler collection) and the window light reflected by the mirror brighten an area that would otherwise be nondescript. *Photo courtesy of Lennon Wallpaper Company, Joliet, Illinois.* (top)

The French *lit clos* (enclosed bed) is clearly defined by a pair of wallcoverings with Oriental overtones, ''Garden of the East'' and ''Garden Print'' from Birge's Oleg Cassini collection. The bed is created by removing closet doors and building in a plywood platform for the mattress (accessible by lacquered stools). This room-within-a-room is then enclosed with shirred-on-the-rod curtains to create a sofa that doubles as a guest bed. The wicker chest stores linens and blankets. Design: Dorothy Wyeth Dobbins. Photography: Photographic House, Inc. *Photo courtesy of Birge Wallcoverings, Inc.* (right)

Colonial elegance returns with the coordinating ''Bridgeport'' and ''Cheltenham'' patterns from Lennon's Colonial Sampler collection. The painted molding strip accentuates the distinction between the two complementary patterns. *Photo courtesy of Lennon Wallpaper Company, Joliet, Illinois.* (top left)

The muted hues of Lennon's ''Essex'' pattern combine with the antique furniture to give this ''correspondence corner'' a touch of eighteenth-century serenity. From the Colonial Sampler collection. *Photo courtesy of Lennon Wallpaper Company, Joliet, Illinois.* (top right)

The look of an old-fashioned screened porch is re-created with white wicker and ''English Garden'' wallpaper, designed for Wall-Tex by Diane von Furstenberg. *Photo courtesy of Columbus Coated Fabrics, Chemical Division, Borden Inc.* (left)

A different kind of wallcovering...

An alternative to the various paper wallcoverings is offered by natural cork available in sheet form, like Armstrong's ''Morena'' pattern shown here. Designs are created by applying overlays of cork to a colored paper base. The base color showing through accentuates the texture of the natural cork to produce interesting visual effects. In some designs, contrasting shades of cork are arranged to create striped, checked, and interlocking effects. Cork wallcovering is also available in designs hand-cut to represent Spanish and American Indian tiles. A unique effect is created when fine-line illustrations are rendered on cork, as in Armstrong's Graphic collection. *Photo courtesy of Armstrong Cork Company*. (right)

And a new way of painting...

A new wall-decorating technique, ''design painting,'' combines the patterned look of wallpaper with the ease of painting. The wall must first be painted with ordinary flat wall paint (oil or latex) in the usual manner, and allowed to dry to the touch. The design can then be applied using the sponge paint applicator and design roller. First the sponge is saturated with paint and then the design roller is mounted on top of the sponge in rolling contact. As the design roller is applied to the wall in parallel strokes, paint is transferred from the sponge to the design roller and from the design roller to the wall to imprint the pattern. A free catalog of patterns can be obtained by writing to Rollerwall, Inc., P.O. Box 757, Silver Spring, Maryland 20901. *Photo courtesy of Rollerwall, Inc.* (left)

PAINTING

Of all the home improvement products on the market, probably none change as rapidly as paint products and tools. Manufacturers are constantly introducing new paint materials which are longer wearing, easier to apply, or flatter or glossier than last year's products. Because of these continual improvements in paint products, only a general description of the major product lines and their application is provided. For the most current information and good advice, consult your paint dealer before undertaking a job. Most paint dealers are very knowledgeable about their wares.

Whether you are painting a room or a house, it's not a job you want to do again soon. The best way to make a paint job last is to prepare the surface properly and to use good quality paints.

As a general rule, the quality of paint is related to its cost. The better the quality, the higher the cost. However, a high priced paint is not necessarily a high quality paint. Therefore, ask your paint dealer for recommendations – and get the best quality paint that you can afford.

After you have selected a finish, read the label on the container for instructions regarding surface preparation, primers, sealers and methods of application. Always make sure that the product you use to finish a surface is compatible with any other materials such as stains, primers or sealers which are on the surface. Also, it is generally recommended that all paint materials used on a surface be from the same manufacturer. This practice will ensure that each surface coating is most compatible with the next.

There are several major families of paint products. They can be grouped as follows:

1. Latex paints and stains. Other names are acrylic or vinyl paint. They are water soluble. They are the easiest to apply and clean up of all paints. Latex paints are fast drying. Many are relatively odorless. They can be applied to damp surfaces. Finishes range from flat to semi-gloss.

2. Oil-based paints and stains. Many of the oil-based paints are also known as alkyds. These paints are soluble with paint thinner. They are generally slow drying in comparison to latex paints. Finishes range from flat to gloss. The semi-gloss and gloss finishes are very resistant to rubbing wear.

---

3. Polyurethanes. These paints are also called urethanes, synthetic or plastic paints. They are soluble with paint thinner. Colored or clear polyurethanes are available. Polyurethanes are more resistant to rubbing and chipping than other paints. Finishes range from satin to gloss.

4. Clear finishes. These finishes are generally used on bare wood or stained wood to allow grain to show.

   a. Varnishes are resin based. They are resistant to rubbing. Spar varnishes are the most weather resistant of the clear finishes. Varnishes are soluble with paint thinner. Finishes range from satin to gloss.

   b. Lacquers are extremely fast drying and flammable. They are soluble with special lacquer thinners. Because of their fast drying time, lacquers can be difficult for inexperienced persons to handle. If new lacquer is brushed onto partly dried lacquer, the partly dried lacquer can be pulled from the surface, thus ruining the finish. Drying can be slowed by adding lacquer thinner. Finish is gloss.

   c. Polyurethanes. See Item 3 above.

   d. Shellac is a natural gum based finish. It is soluble with shellac thinner (alcohol). It is easy to apply without leaving brush marks. If the shellac finish becomes damaged by wear or water, it is easy to repair. Shellac is easily damaged by water. If water is left on the finish, it can leave white spots. Finish is gloss.

The following charts show common applications of these products to various interior and exterior surfaces.

Interior Finishes

Surface \ Finish	Latex Wall Paint	Latex Enamel	Oil-Based Enamel	Spray Enamel	Latex Floor Enamel	Oil-Based Floor Enamel	Latex Masonry Paint	Portland Cement Paint	Stains	Varnishes	Polyurethanes	Penetrating Wood Finishes	Shellac	Lacquer	Antique Kits
Cabinets and Furniture															
Metal			X	X											
Wood			X	X					X[6]	X	X	X	X	X	X
Ceilings															
Acoustical, tile	X[1]														
Acoustical, spray-on	X														
Plaster or wallboard															
Bathroom, kitchen		X	X												
Other	X														
Doors															
Metal		X	X												
Wood		X	X						X[6]	X	X	X	X	X	
Floors															
Concrete					X[4]	X[4]									
Wood					X	X			X[6]	X	X	X	X	X	
Trim															
Metal		X	X												
Plastic		X	X												
Wood		X	X						X[6]	X	X	X	X	X	
Walls															
Concrete	X[2]	X[2]					X[2]	X[2,5]							
Plaster or wallboard															
Bathroom, kitchen		X	X												
Other	X														
Wood	X[3]	X	X						X[6]	X	X		X	X	

1. Paint will reduce the sound deadening quality of acoustical tile. A special acoustical paint is available for acoustical tile. It is designed to be non-bridging so that the sound deadening perforations in the tile are not covered by the paint.

2. Latex paint is especially resistant to the alkalis in concrete. It also does an excellent job of filling the rough pores in masonry blocks. It will adhere well to masonry surfaces. Be sure to prepare new or old concrete surfaces as directed on label.

3. Latex wall paint has a very flat finish which may not look as good on wood surfaces as latex enamels or oil-based enamels.

4. Latex floor enamel is not as resistant to rubbing as oil-based floor enamel. It is not recommended for use on garage floors where it will be rubbed by tires. Be sure to prepare concrete floors as directed on label.

5. May be difficult to cover with other paints at a future time.

6. If clear finish is to be used, make sure stain is compatible with finish.

Exterior Finishes

Surface \ Finish	Exterior Latex House Paint	Exterior Oil-Based House Paint	Latex Trim Enamel	Exterior Oil-Based Trim Enamel	Latex Floor Enamel	Oil-Based Porch and Floor Enamel	Exterior Latex Stain	Exterior Oil-Based Stain	Exterior Polyurethane Varnish	Exterior Varnish (Spar Varnish)	Latex Masonry Paint	Portland Cement Paint
Ceilings and Eaves												
Metal	X	X										
Wood	X	X					X	X				
Decks and Floors												
Brick					X	X						
Concrete					X[1,4]	X[1,4]						
Stone					X	X						
Wood					X	X		X	X[5]	X		
Doors												
Metal		X[3]	X	X								
Wood		X[3]	X	X					X[5]	X		
Fences												
Brick	X										X	
Concrete	X[1]										X[1]	X[1,6]
Metal	X	X										
Wood	X						X	X				
Furniture												
Metal				X								
Wood				X				X	X	X		
Gutters												
Metal	X	X										
Wood	X	X					X	X	X			
Shutters and Trim												
Metal	X[2]	X	X	X								
Wood	X[2]	X	X	X			X	X	X	X		
Walls												
Asbestos shingles	X										X	
Brick	X[1]										X[1]	
Concrete	X[1]										X[1]	X[1,6]
Metal	X	X									X	
Stucco	X											
Wood shingles	X						X	X				
Wood siding	X	X								X		

1. Latex paint is especially resistant to the alkalis in concrete. It also does an excellent job of filling the rough pores in masonry blocks. It will adhere well to masonry surfaces. Be sure to prepare new or old concrete surfaces as directed on label.

2. Latex house paint has a flat finish. The semi-gloss or gloss finish provided by a trim enamel may be more attractive.

3. Consider using a harder surface trim enamel for doors. Doors receive much rubbing and chipping wear which is best resisted by trim enamels.

4. Latex floor enamel is not as resistant to rubbing as oil-based floor enamel. It is not recommended for use in driveway areas where it will be rubbed by tires. Be sure to prepare concrete floors as directed on label.

5. Spar varnish is the most resistant of clear finishes to weather. Therefore, consider using spar varnish rather than polyurethanes on decks or large, exposed doors such as garage doors.

6. May be difficult to cover with other paints at a future time.

ESTIMATING PAINT NEEDS

The amount of surface a paint will cover depends upon:

- Paint formula (Label on container shows maximum area of coverage. Do not try to cover more area. Paint will be spread too thinly.)
- Type of surface to be covered (smooth clapboard, rough stucco)
- Whether primer, sealer or undercoat has been applied to the surface
- Type of tool used to apply the paint

Knowing the surface area and the type of surface (masonry, finished wood, wallboard, etc.), the paint dealer can tell you the amount of paint you need. Try to have some paint left over for later touch-up and repair.

Most surface areas can be divided into rectangles and triangles. For as close an estimate as possible total area of each rectangle and triangle should be determined in square feet, and these quantites added to give total surface area.

When estimating paint for surfaces interrupted by doors, windows, or fireplaces, make your estimate for the entire surface. Then make your estimate for each door, window, and fireplace and subtract these areas from the estimate for the entire surface.

PAINTING TOOLS

This section describes the different types of paintbrushes, rollers, and other tools and supplies used for painting. Instructions for selecting and using brushes, rollers and other paint spreaders are also included.

▶ Paintbrushes

Most paint jobs around the home can be done with the following brushes:

- One-inch brush [1] is used to paint small, hard to reach places.
- Two-inch brush [2] with chiseled edge is used to paint with enamel and varnish.
- Angled sash brush [3] is used to paint window frames, moldings, and other narrow surfaces.
- Four-inch brush [4] is used to paint large, flat surfaces.

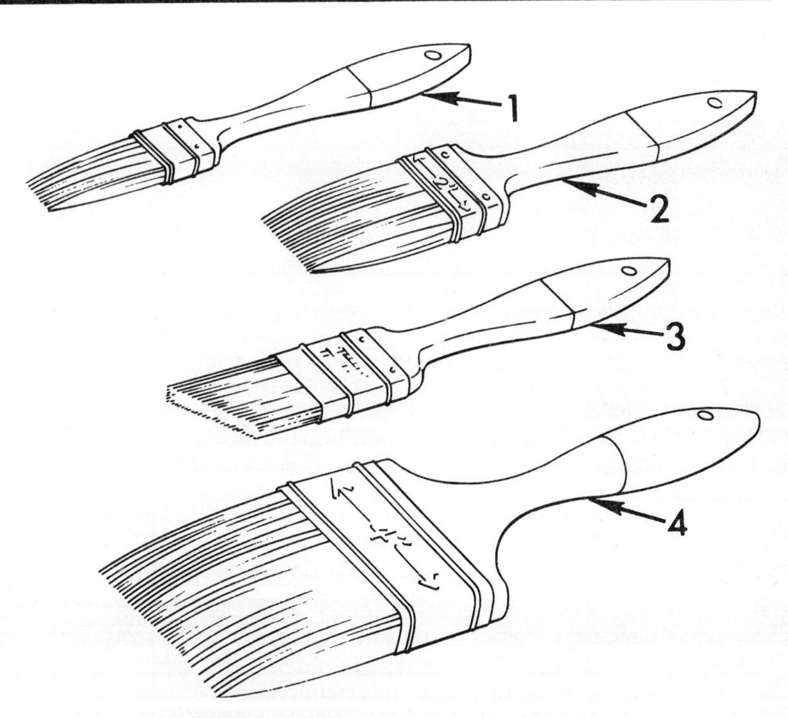

▶ Selecting a Brush

For any painting job, always use a good paintbrush. Properly used and cared for, it will give you many years of service.

A good paintbrush establishes the ease of painting and the quality of the finished job.

A good brush holds more paint, controls dripping, and applies paint more smoothly.

PAINTING TOOLS

Selecting a Brush

Paintbrushes are made with either of the following two types of bristles:

- Natural bristle brushes are made from animal hair. Because natural bristles absorb water, they should not be used with water soluble paints. They are used for applying varnish, shellac, lacquer, polyurethane finishes, and oil-based paints.

- Synthetic bristle brushes are made from man-made fibers such as nylon. Because of their ability to resist water absorption, synthetic bristle brushes are used with paints that use water for cleanup.

  Synthetic bristles, particularly nylon, can now be made which look and work like natural bristles. A good nylon bristle brush can be used to apply oil-based finishes as well as water soluble finishes with good results.

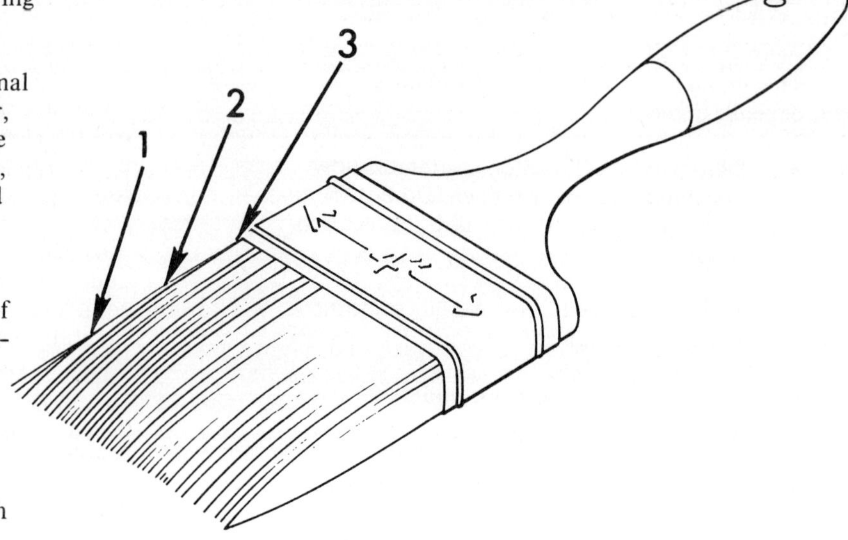

1. Check that bristles [2] are flagged. Flagged bristles have split ends [1] which hold more paint and apply paint more smoothly.

2. Check that bristles [2] are tapered. Tapered bristles are thicker at the base [3] than at the tip which helps paint flow more smoothly.

Selecting a Brush

3. Check brush for fullness. Press bristles [1] against hand. Check that bristles are full and springy.

4. Check that divider [2] in brush setting [3] is not so large as to cause a hollow space in center of brush. If divider is too large, bristles will not feel thick and brush will hold less paint.

5. Check that bristles [1] are not all the same length. Pull hand over bristles. Check that shorter bristles spring up.

6. Check that bristles [1] are firmly held against setting [3]. Bristles should be attached to setting with epoxy glue or other bond. Nails should be used only to hold ferrule [4] to handle [5].

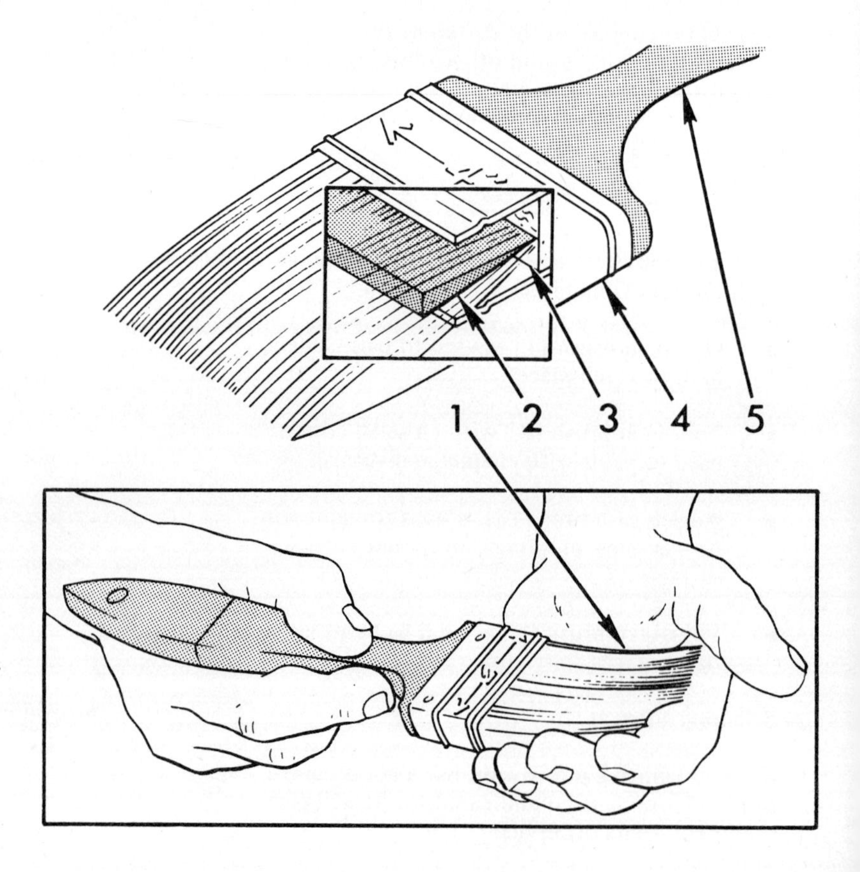

▶ Using a Brush

Be sure to read entire procedure before beginning to paint.

1. Prepare paintbrush by lightly moistening bristles in solvent used for cleanup. Squeeze solvent into bristles.

2. Remove solvent from bristles with clean, absorbent cloth.

Applying paint to a surface consists of four steps:

- Filling the brush with paint
- Placing the paint on the surface
- Spreading the paint evenly on the surface
- Smoothing the paint to remove visible brush marks and overlap.

When dipping the brush into the paint, do not dip bristles into the paint more than one-half the length of the bristles.

The first time the brush is dipped into the paint, gently stir the paint with the brush to spread the bristles slightly.

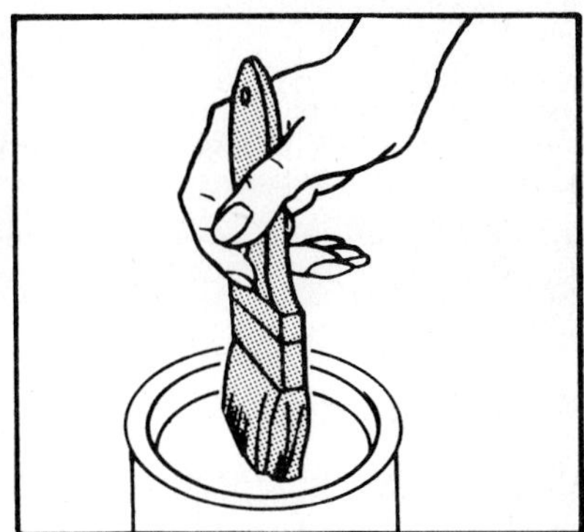

3. Dip brush into paint until bristles are one-half their length into the paint. Lift brush straight up from paint. Allow excess paint to drip off bristles.

Some experts recommend that you do not pull bristles across edge of bucket to remove excess paint. This makes it difficult to keep brush clean and may separate bristles into clusters so that it is difficult to smooth brush marks from painted surface.

4. Gently tap bristles against inside of bucket to remove excess paint.

Using a Brush

Paint must be applied to unpainted surface [2] and spread into freshly painted surface [1].

CAUTION

When placing paint on surface, be careful to avoid spattering paint.

5. Place paint on unpainted surface at several places [3] by gently slapping surface with flat sides of brush.

When spreading paint, hold brush at approximately a 45° angle. Press brush firmly enough to make all bristle ends touch surface, and pull brush across surface.

6. Spread paint over surface to blend marks [3] together.

If wood is being painted, finish strokes must be in direction of wood grain.

7. Smooth paint with long, light brush strokes.

Proper brushing technique is important to avoid visible overlaps between painted areas. Overlaps

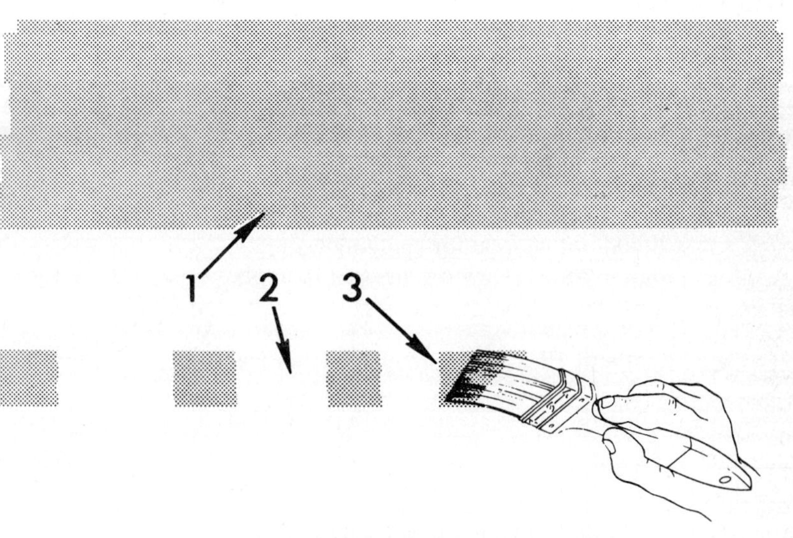

are caused by paint being spread too thickly at edges.

To avoid visible overlaps, paint must be spread more thinly at edges than at centers of freshly painted areas. This technique is called "feathering".

To spread paint more thinly at edges, begin gradually lifting brush from surface as you approach edges.

PAINTING TOOLS

▶ Rollers

Paint rollers are best used to paint large areas. They apply paint much faster than a paintbrush while still giving a smooth, uniform coat of paint.

Most paint jobs around the home can be done with the following rollers:

- Three-inch trim roller [1] is used to paint window frames, moldings, and other narrow surfaces.
- Three-inch diameter corner roller [2] is a wedge-shaped roller used to paint corners.
- Seven- or nine-inch roller [3] is used to paint large, flat surfaces. A nine-inch roller is recommended because most persons can easily handle this length.

▶ Selecting a Roller

Many kinds of roller covers [4] are available. Different jobs require different covers. For example, a cover with a short nap is used for smooth surfaces such as interior walls and ceilings.

A long nap is needed for rough surfaces such as exterior stucco. Your paint dealer will help you select the best roller cover [4] for your job.

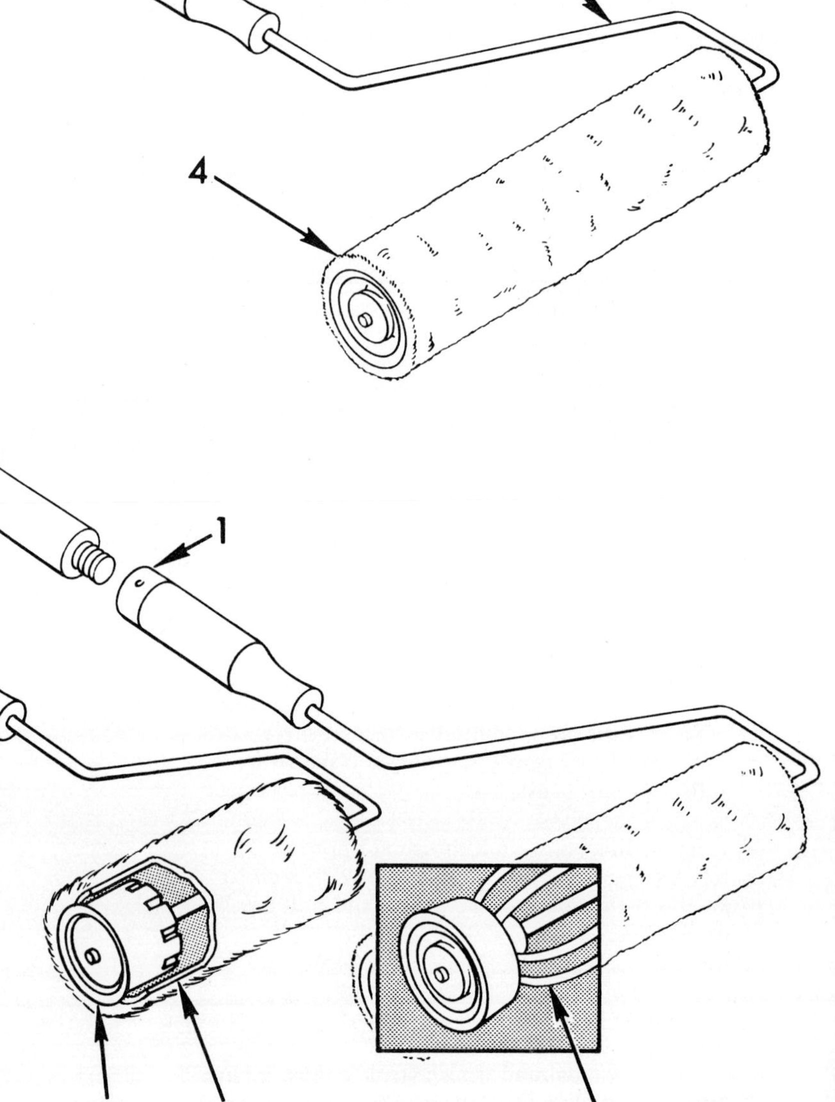

Selecting a Roller

Roller covers are made with either of the following two types of naps:

- Natural fiber covers are made from wool or mohair. Because they absorb water, they should not be used for applying water soluble paint. They can be used to apply oil-based paints.
- Synthetic fiber covers are made from man-made fibers such as rayon and nylon. Synthetic covers are used with paints that use water for cleanup. They also work well with oil-based paints.

1. Check that roller has cage frame [4] or nylon end caps [2].
2. Check that roller cover core [3] is made of plastic-covered cardboard.

The texture of the surface to be painted determines the length of the nap.

3. For smooth surfaces, use roller cover with short nap (1/4-inch). For rough surfaces, use roller cover with long nap (3/8-inch or more).
4. Check that handle [1] is threaded to allow attaching extension handle.

▶ Roller Accessories

The following tools are needed if using a roller:

- An extension handle [1] to reach high places without using a ladder. A 24-inch extension handle is adequate for most interior jobs. This length is easy for most persons to handle.
- A roller tray [2] to load roller. Roller tray must be wider than roller. The deeper the roller tray, the less often it will have to be filled.

Cover the tray with a plastic bag so that at cleanup time you simply remove and discard the bag.

- A plastic or metal grill [3] to help load a roller evenly.

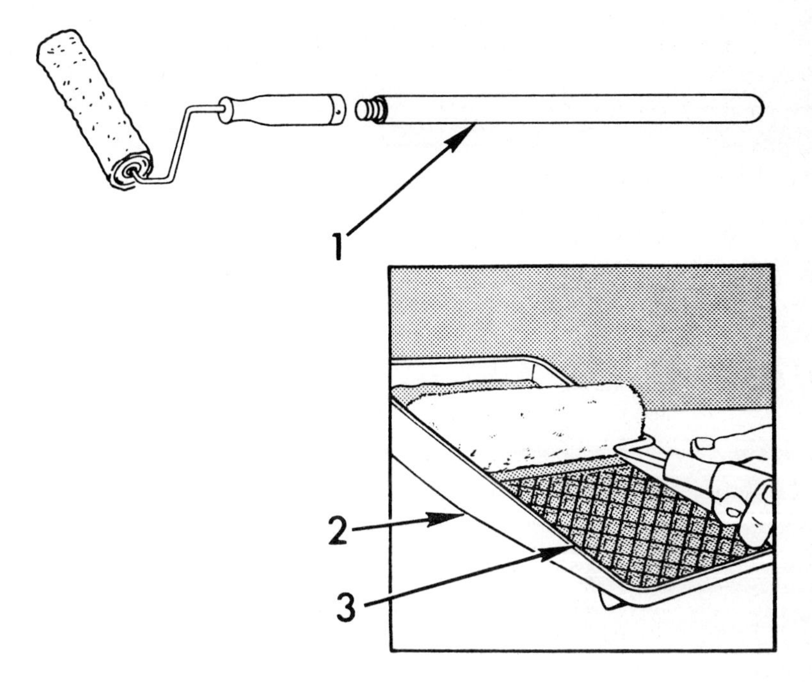

▶ Using a Roller

Be sure to read entire procedure before beginning to paint.

1. Prepare roller by lightly moistening roller cover with solvent used for cleanup. Squeeze solvent into nap.
2. Remove solvent from nap with clean, absorbent cloth.

Using a Roller

Applying paint to a surface consists of four steps:

- Filling the roller with paint
- Placing the paint on the surface
- Spreading the paint evenly on the surface
- Smoothing the paint to remove visible roller marks and overlap.

When loading roller with paint, the object is to thoroughly and evenly soak entire roller.

Roller is rolled up tray or grill to distribute paint evenly in roller and to remove excess paint.

3. Dip roller into paint. Pull roller up the roller tray or grill.

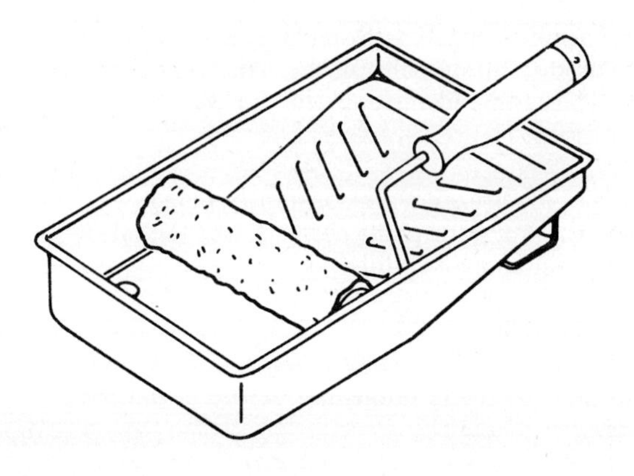

PAINTING TOOLS

Using a Roller

When placing paint on surface, begin with light pressure strokes. Increase pressure as roller places paint on the surface. This technique will result in release of paint evenly on surface.

CAUTION

Paint droplets will be spattered in area of roller if roller is rolled too fast. Speed will vary with type of roller and paint.

4. Beginning with an upward stroke, place paint on surface in an M-shaped pattern [1].

5. Spread paint over surface by rolling roller up and down and back and forth.

If wood is being painted, finish strokes must be in direction of wood grain.

6. Smooth paint evenly over freshly painted area.

Proper rolling technique is important to avoid visible overlaps [3] between painted areas [2,4]. Overlaps are caused by paint being spread too thickly.

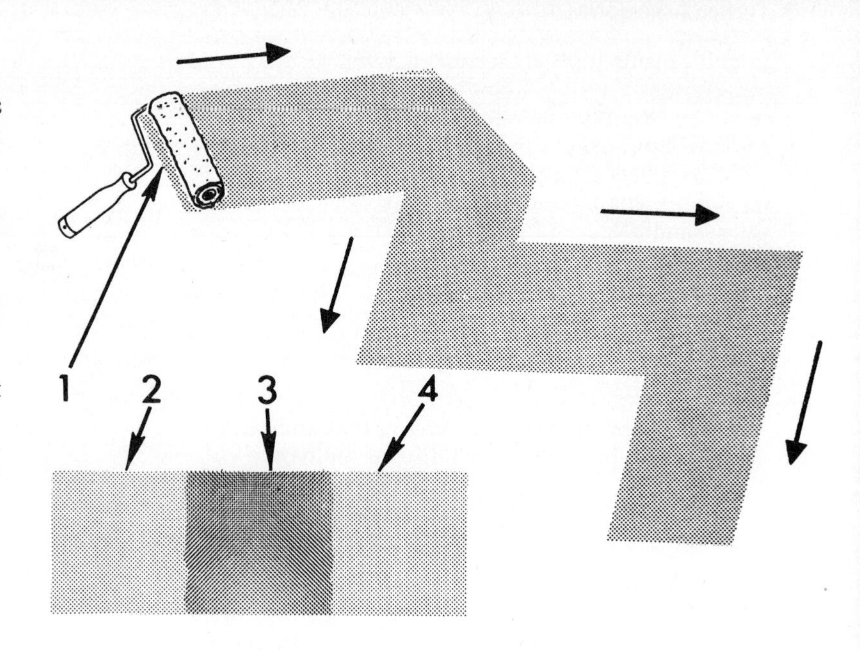

To avoid visible overlaps [3], roll overlap with roller after it has released most of its paint. Roller will pick up and spread excess paint.

▶ Other Paint Spreaders

A smooth pile spreader [2] with guide wheels [1] can be used to paint edges of a surface where the wheels can be rolled along a guide. These spreaders are also available without wheels.

A paint applicator [3[with replacement pads [4] can be used to apply paint in the same way as a brush. Pads come in various sizes for different types of jobs.

New types of spreaders are constantly being developed. Ask your paint dealer about the latest type of paint spreading aids.

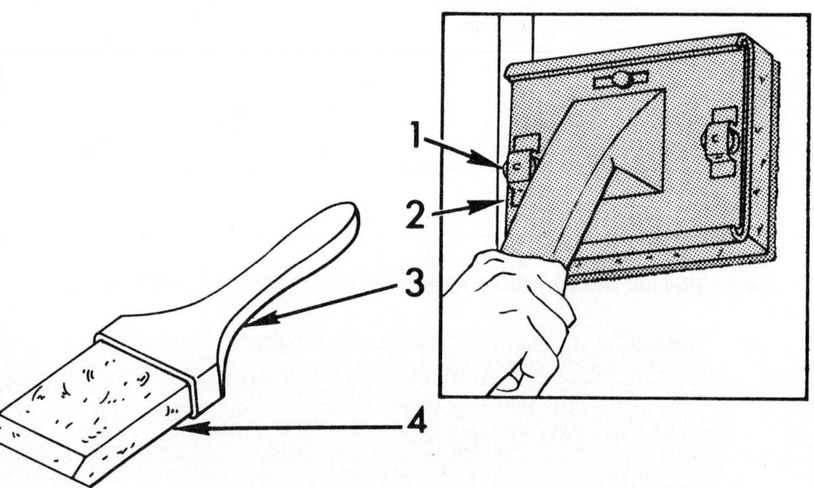

▶ Using a Spreader

Be sure to read entire procedure before beginning to paint.

1. Prepare spreader by lightly moistening spreader with solvent used for cleanup. Squeeze solvent into nap.

2. Remove solvent from nap with clean, absorbent cloth.

Applying paint to a surface consists of four steps:

- Filling the spreader with paint
- Placing the paint on the surface
- Spreading the paint evenly on the surface
- Smoothing the paint to remove visible spreader marks and overlap.

Using a Spreader

When dipping spreader [1] into paint, be sure to dip only the pad [2] into the paint.

3. Dip spreader [1] into paint. Lift spreader straight up from paint. Allow excess paint to drip off pad [2].

4. Using one stroke with spreader [3], place paint on surface.

5. Spread paint over surface by using continuous strokes all in same direction.

If wood is being painted, finish strokes must be in direction of wood grain.

6. Smooth paint with long, light spreader strokes.

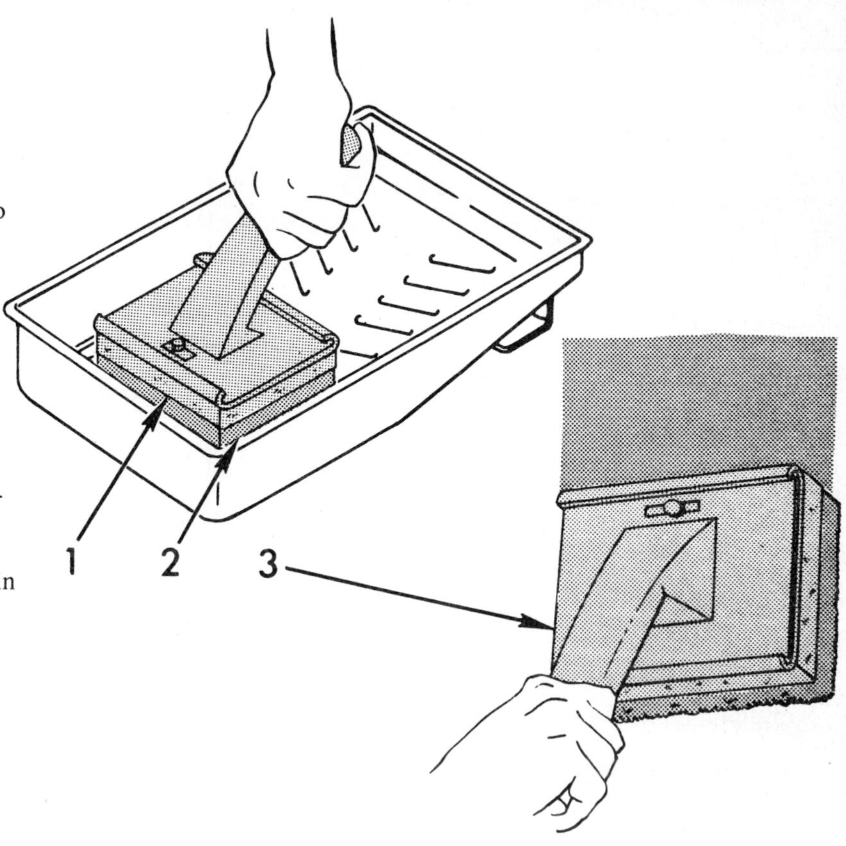

▶ Other Tools and Supplies

The following tools and supplies are needed in addition to paintbrushes, rollers, and spreaders:

- Small bucket [1] or pail to hold paint being applied.
- Quality masking tape [2] to protect unpainted surfaces.
- A stepladder [3] or stool to reach high places. If painting ceiling or exterior, two ladders and scaffold planks should be used.
- Drop cloths [4] to protect the floor and furniture not removed from a room. If painting exterior surfaces, drop cloths are used to protect bushes and shrubs.
- Clean, absorbent rags.

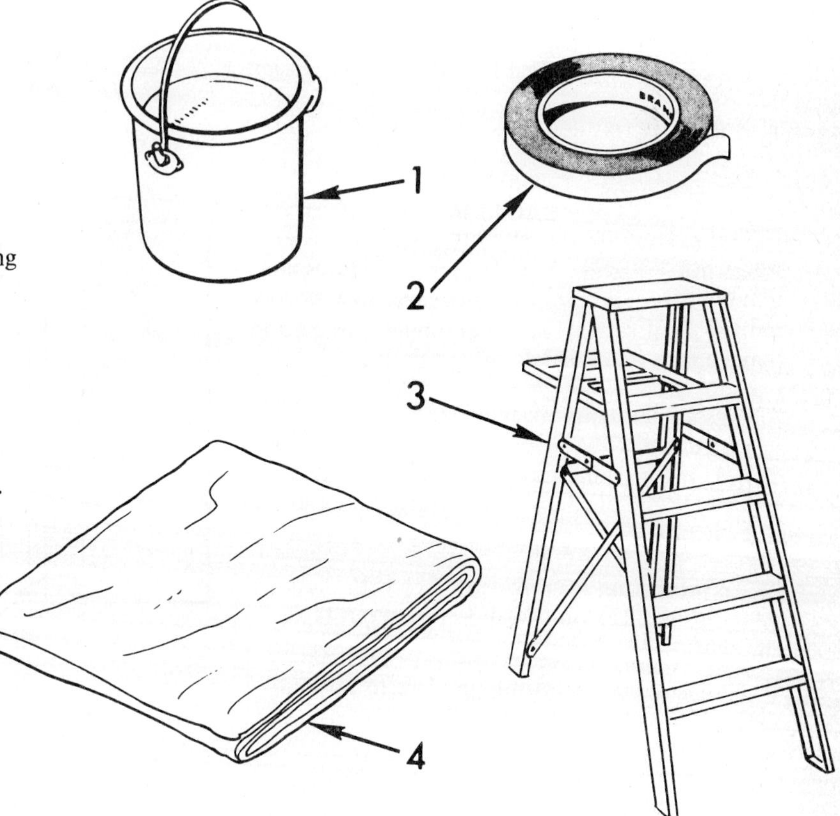

Instructions in this section tell how to prepare the work area and surface for the application of interior and exterior paints. How to mix paint is also described.

▶ Interior Preparation

A completely empty room is the best work area. Furniture that cannot be removed should be moved away from work areas.

A completely bare surface is the best for painting. Curtains, pictures, and wall fixtures such as curtain rods and switch covers should be removed.

Drop cloths should be placed on the floor and on any furniture not removed from the room.

Applying paint over old wallpaper is not recommended. When the paint is applied, the wallpaper will become wet and may pull loose from the surface. Also, if old wallpaper is painted, it will be more difficult to remove the wallpaper in the future.

If surface is wallpapered, remove wallpaper before continuing.

Interior Preparation

Doors should be removed to make painting easier.

1. Remove fixtures from doors, windows, and cabinets.

CAUTION

When using masking tape to protect fixtures and windowpanes, be sure to remove masking tape as soon as paint sets. If in direct sunlight, be sure to remove tape immediately after painting.

Always use a quality masking tape.

2. Using masking tape or plastic bags and masking tape, cover all fixtures that are not removed.

When protecting windowpanes [1] with masking tape, allow a 1/16-inch gap [2] between tape and putty.

Be sure to press tape firmly in place to prevent paint from running under tape.

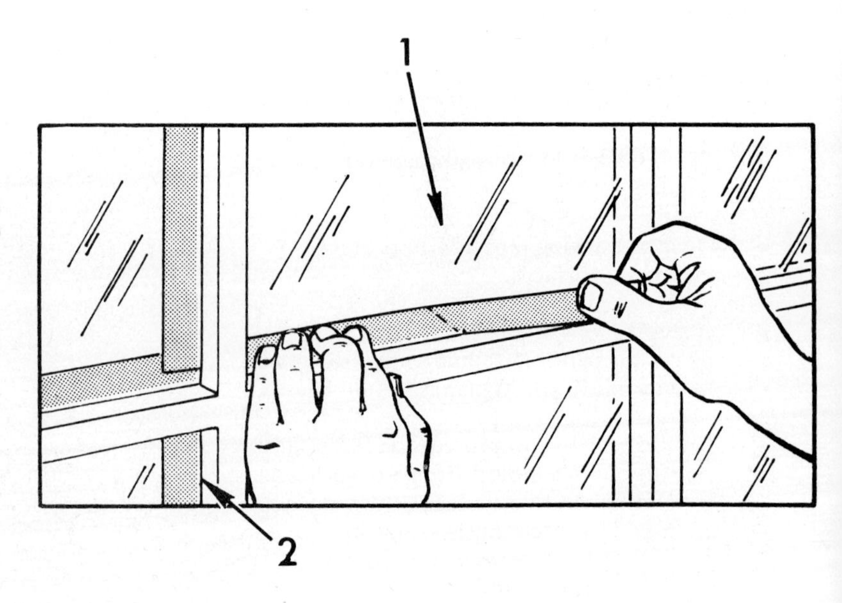

3. Using masking tape, cover edges of windowpanes [1].

Interior Preparation

4. Inspect all surfaces to be painted for holes, cracks or water damage.

If problems are found, repair problems before continuing. Review House Repairs, Page 99.

5. Following manufacturer's instructions, mix a solution of trisodium phosphate (T.S.P.) and water. Thoroughly wash surface. Rinse well with clear water. Allow to dry completely.

Paint will not adhere well to surfaces that are shiny or glossy. If surface is shiny or glossy, continue.

Sandpaper can be used to degloss the surface.

Commercial deglossers are also available to remove the gloss or shine from painted surfaces. Ask a paint dealer about them.

6. Using fine sandpaper or commercial deglosser, remove gloss from painted surface.

▶ Exterior Preparation

A completely bare surface is the easiest surface to paint. Screens, shutters, mailbox, house numbers, and any other accessories should be removed.

1. Remove fixtures from doors and windows.

CAUTION

When using masking tape to protect fixtures and windowpanes, be sure to remove masking tape as soon as paint sets. If in direct sunlight, be sure to remove tape immediately after painting.

Always use a quality masking tape.

2. Using masking tape or plastic bags and masking tape, cover all accessories or fixtures that are not removed.

When protecting windowpanes [1] with masking tape, allow a 1/16-inch gap [2] between tape and putty.

Be sure to press tape firmly in place to prevent paint from running under tape.

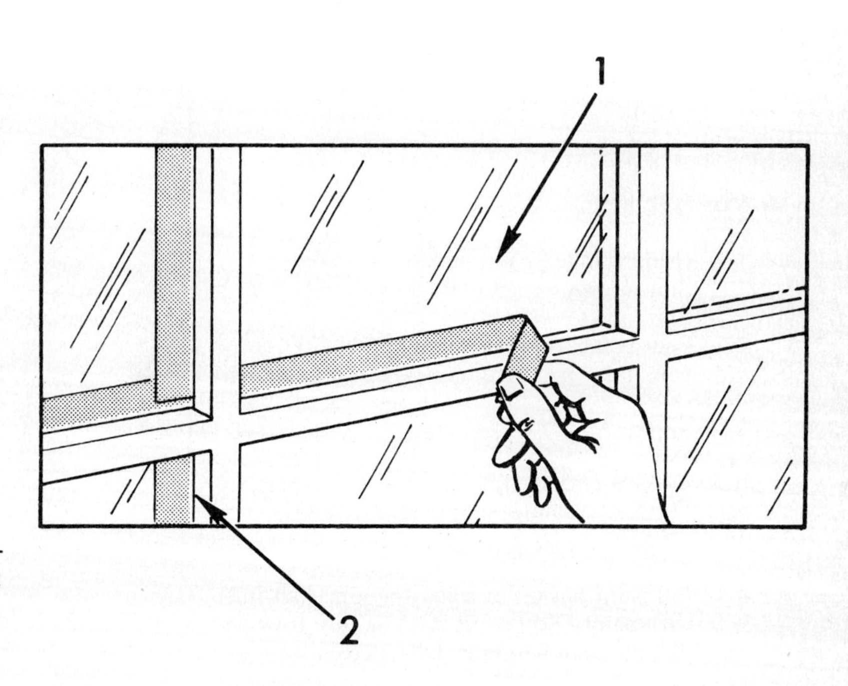

3. Using masking tape, cover edges of windowpanes [1].

Exterior Preparation

4. Inspect all surfaces to be painted for problems.

If problems are found, repair problems before continuing. Review Paint Problems and Repairs, Page 92.

5. Rinse surface well with hose and clear water. Allow to dry completely.

6. Check that surface is clean.

If surface is not clean, continue.

7. Following manufacturer's instructions, mix a solution of trisodium phosphate (T.S.P.) and water. Thoroughly wash surface. Rinse with clear water. Allow to dry completely.

When working on new surfaces, sanding prior to painting improves the finish and adhesion. Old surfaces should be sanded for similar reasons.

Paint will not adhere well to surfaces that are shiny or glossy.

Sandpaper can be used to degloss the surface.

Commercial deglossers are also available to remove the gloss or shine from painted surfaces. Ask a paint dealer about them.

8. Using fine sandpaper or commercial deglosser, remove gloss from painted surfaces.

9. Tie shrubs and bushes away from exterior walls. Cover with drop cloths.

10. Place drop cloths over patios, porches, driveways, and any other surfaces to be protected.

▶ Mixing Paint

1. Using wooden paint paddle, stir paint thoroughly.

2. Pour paint from paint can into bucket or pail. Pour paint from bucket into paint can.

3. Repeat Step 2 three times.

Before filling a roller tray, cover it with a plastic bag. For easy cleanup, simply remove and discard bag.

4. Fill paint bucket or roller tray one-half full with paint. Seal paint can. Be sure to wipe lip [1] clean before installing cover.

▶ Interior Sequence

Instructions in this section describe the sequence to be followed for interior painting.

Interior painting is done in the following sequence with ceilings painted first and walls painted last:

Ceilings, Page 79.
Windows, Page 79.
Doors, Page 82.
Masonry
Baseboards
Cabinets and Cupboards, Page 83.
Walls, Page 83.

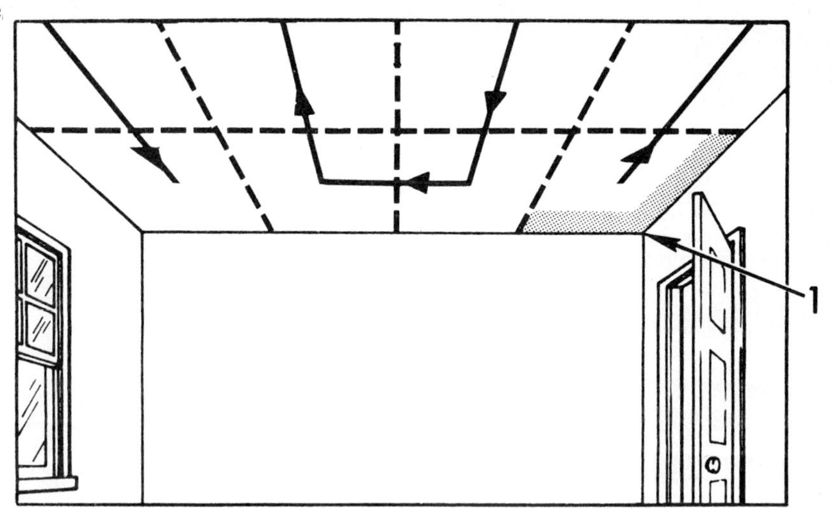

▶ Interior — Ceilings

Ceilings are painted in section of approximately 3 feet by 3 feet.

Painting is begun in a corner of the ceiling. Paint across ceiling in the direction with the shortest distance.

Within each section, first paint a 2-inch strip along ceiling line [1]. Then paint the remainder of the section.

Paint a 2-inch strip around any obstacles before painting within a section.

▶ Interior — Frame Windows

1. Raise inside sash to approximately 3/4-open position. Lower outside sash to approximately 3/4-open position.

When painting window, be sure to apply paint 1/16-inch onto glass to make an airtight seal.

2. Paint inside sash as follows:

   Horizontal muntins [1]
   Vertical muntins [2]
   Top sash [3]
   Two vertical sashes [4]
   Bottom sash and bottom edge [5]

3. Paint outside sash as follows:

   Horizontal muntins [6]
   Vertical muntins [7]
   Two vertical sashes [8]
   Bottom sash [9]

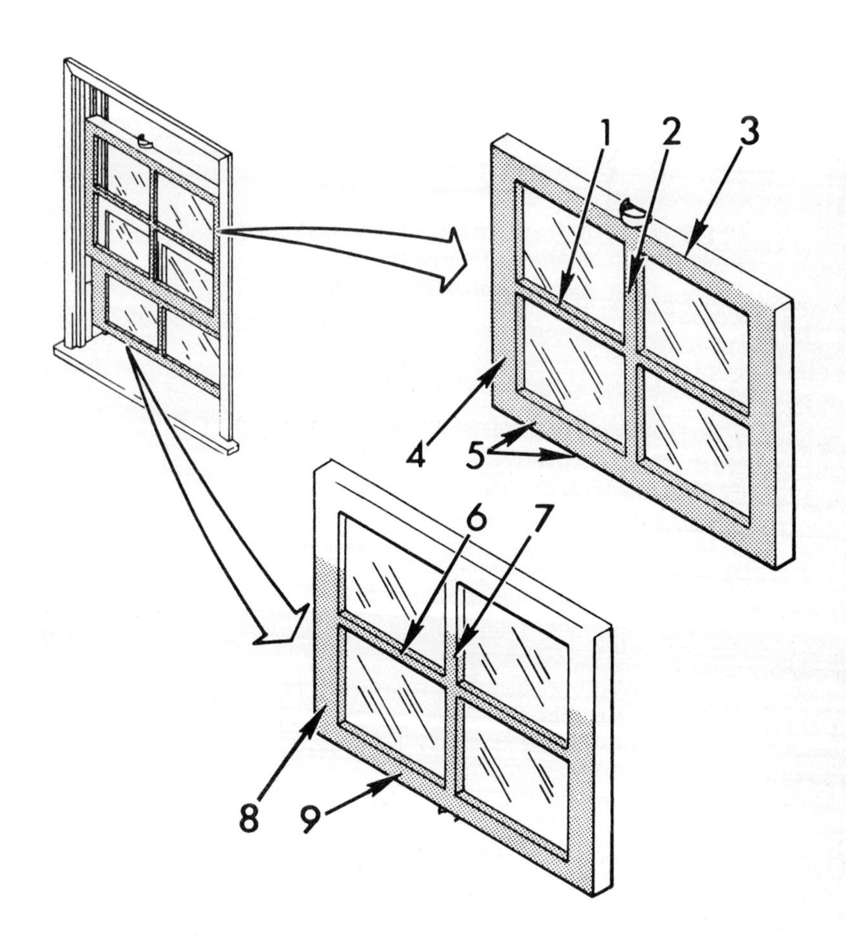

SEQUENCE OF PAINTING

Interior – Frame Windows

4. Lower inside sash to approximately 1 inch from fully closed position. Raise outside sash to approximately 1 inch from fully closed position.

When painting window, be sure to apply paint 1/16-inch onto glass to make an airtight seal.

5. Paint remaining surfaces of outside sash as follows:

   Horizontal muntins [1]
   Vertical muntins [2]
   Top sash [3]
   Two vertical sashes [4]

6. Paint top edge [5] of inside sash.

7. Paint frame as follows:

   Head casing [6]
   Two side casings [7]
   Stool [8]
   Apron [9]

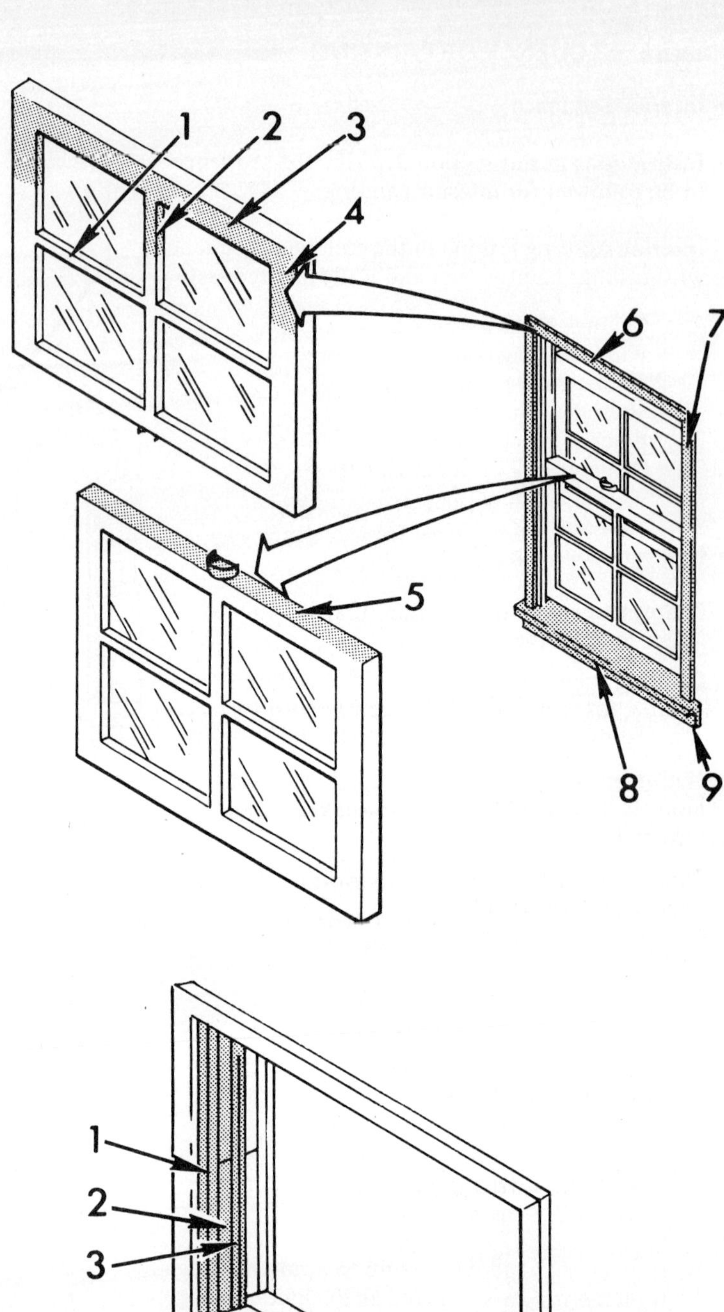

Interior – Frame Windows

CAUTION

To prevent them from sticking, inside and outside sashes should be moved frequently while paint is drying. When moving sashes, do not close them completely.

8. Allow paint to dry completely.

When painting window jambs, only the surfaces which are seen when the window is fully closed are to be painted.

Be careful not to allow paint to run down into grooves on lower half of jamb.

9. Paint top half of jamb as follows:

   All three sides of forward stop [1]
   Channel [2]
   Two visible sides of bind stop [3]

10. Paint lower half of jamb as follows:

   Two visible sides of forward stop [1]

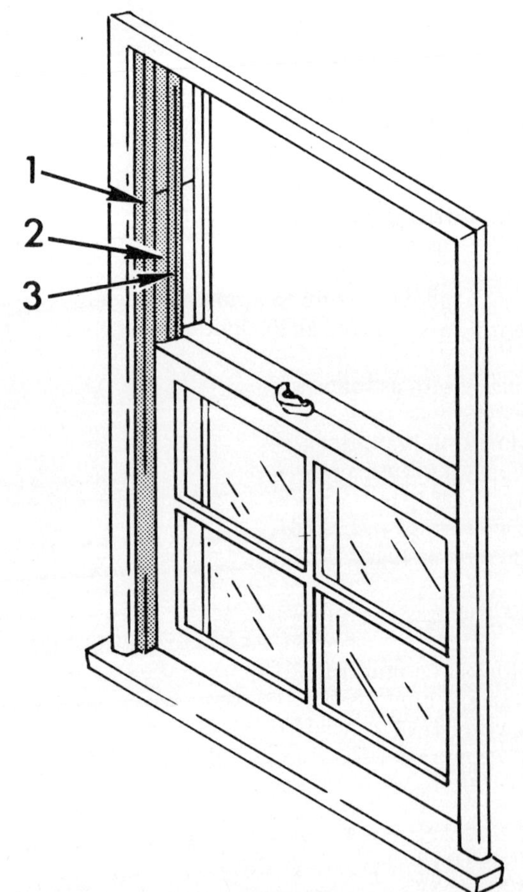

Interior – Frame Windows

11. Allow paint to dry completely.

12. Raise inside sash to fully open position.
    Raise outside sash to fully closed position.

13. Using beeswax or paraffin, rub lower half of jamb of inside sash as follows:

    Inside surface of forward stop [1]
    Channel [2]
    Bind stop [3]

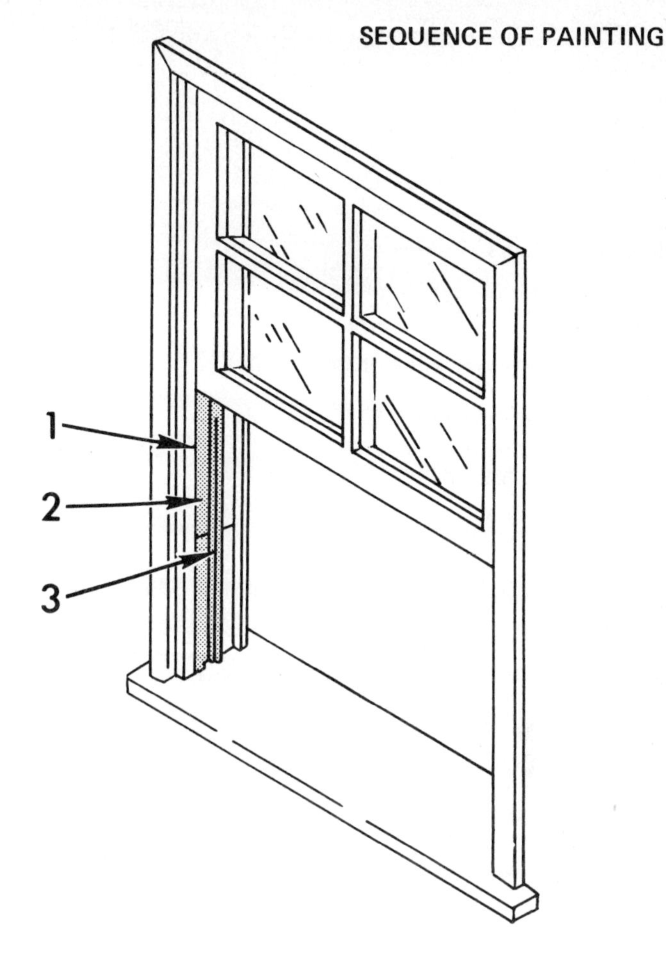

▶ **Interior – Casement Windows**

1. Open window to fully open position.

When painting window frame, be sure to apply paint 1/16-inch onto glass to make an airtight seal.

When painting window, paint edges [1] of strips facing the inside only.

2. Paint window as follows:

    Horizontal strips [2]
    Vertical strips [3]

3. Paint frame as follows:

    Horizontals [4]
    Verticals [5]

4. If window has stool and apron, paint stool and apron as follows:

    Stool [6]
    Apron [7]

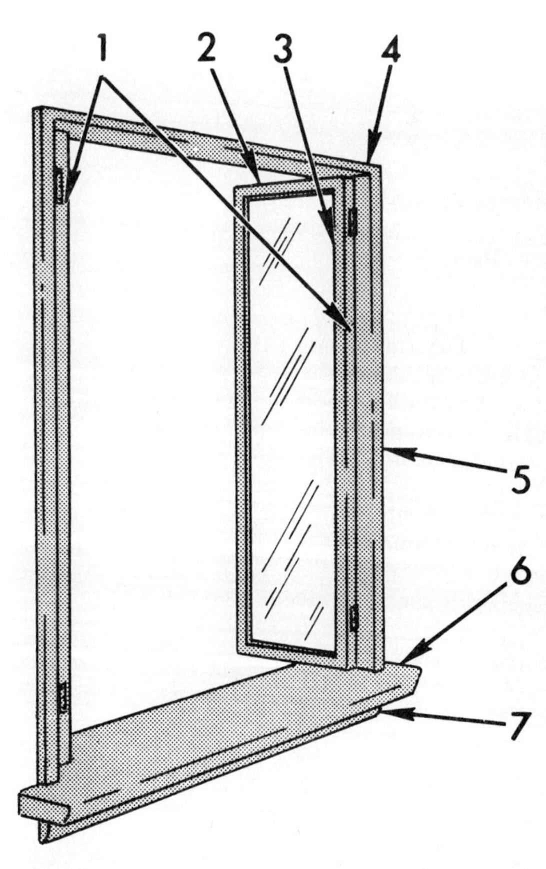

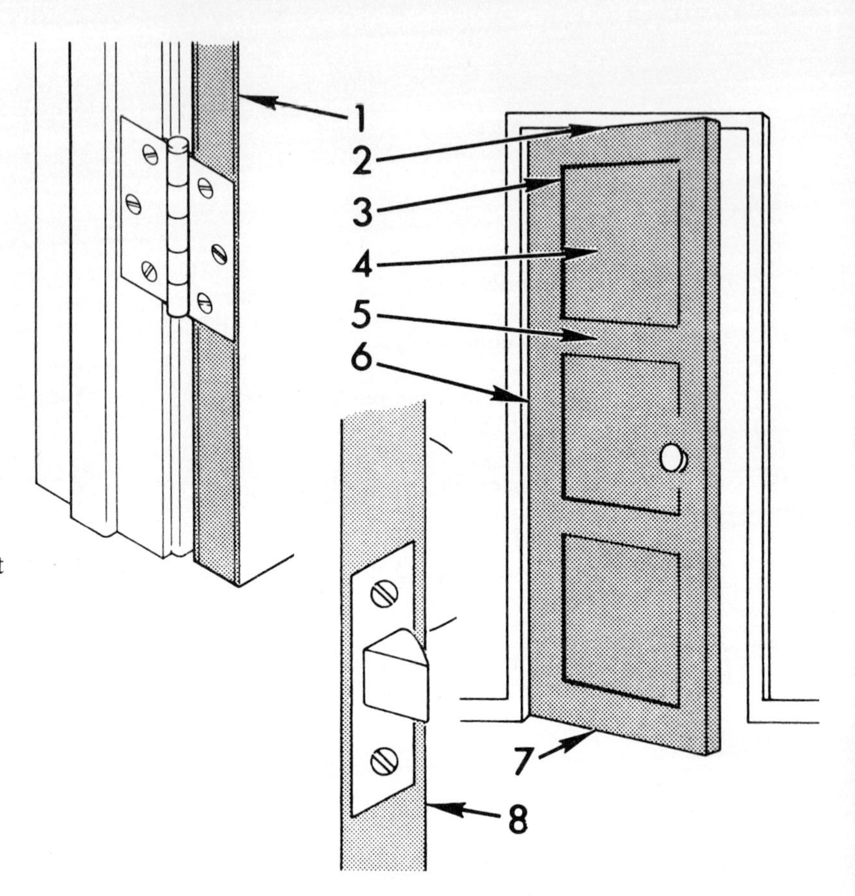

▶ **Interior – Doors**

1. Paint door as follows:

   Inside edges [3] of panel cavities
   Panels [4]
   Horizontal strips [5]
   Vertical strips [6]
   Top edge [2]

2. If door is removed, paint bottom edge [7].

3. If door opens into room being painted, paint latch edge [8].

4. If door does not open into room being painted, paint hinge edge [1].

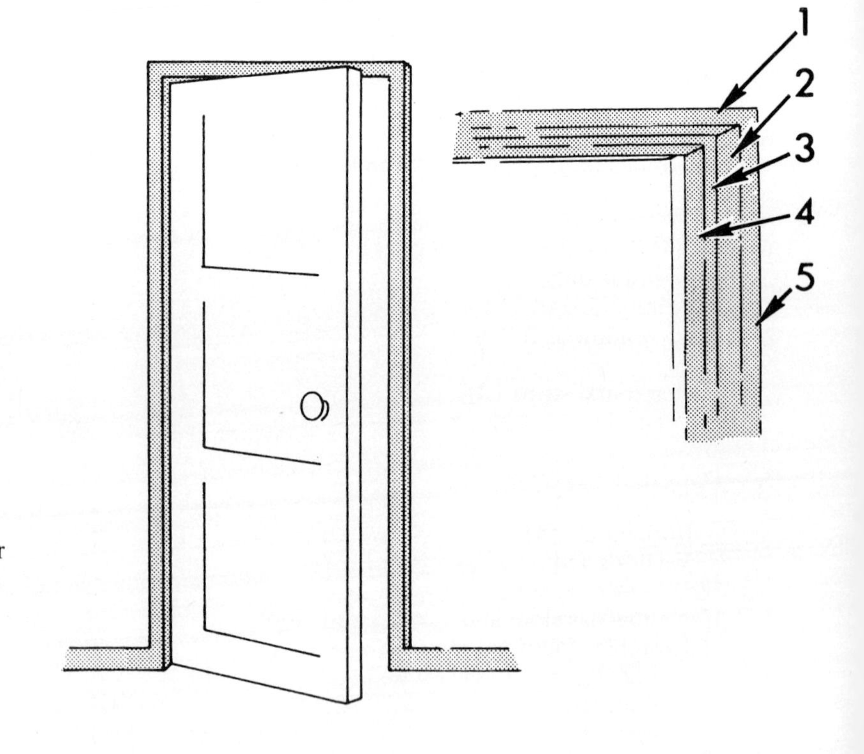

Interior – Doors

5. Paint frame as follows:

   Top frame [1]
   Two side frames [5]

If door opens into room being painted, perform Step 7. If door does not open into room being painted, perform Step 6.

6. Paint jamb [2]. Paint two sides [3,4] of door stop.

7. Paint jamb [2]. Paint door side [3] of door stop.

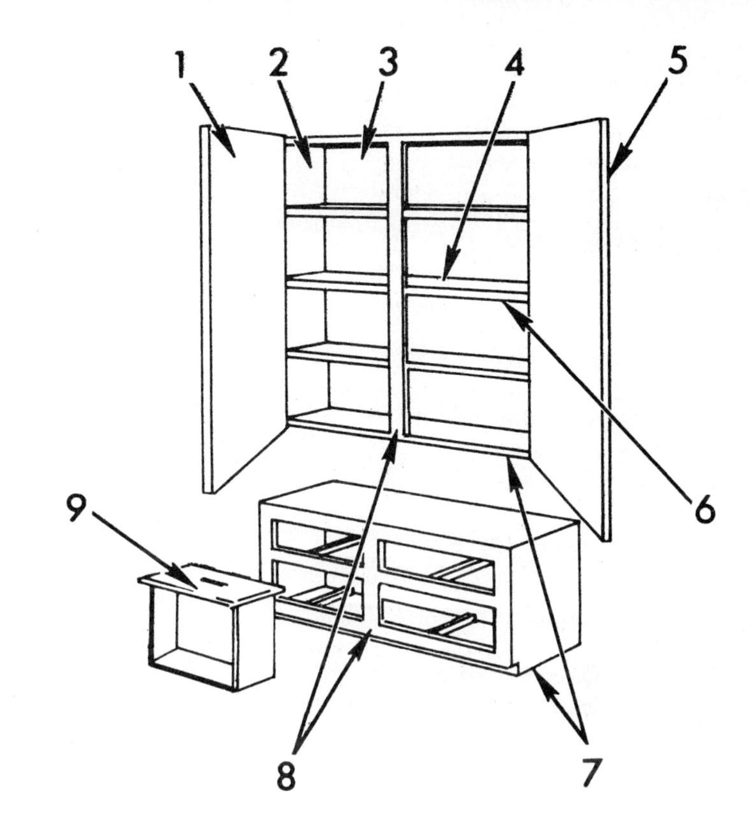

▶ **Interior – Cabinets and Cupboards**

1. Paint cabinets and cupboards as follows:

   Inside surface [1] of doors
   Inside walls [2]
   Back wall [3]
   Shelf tops [4] and edges
   Outside surface [5] of doors
   Shelf bottoms [6]
   Outside surface [7,8] of cabinets
   Drawers [9]

▶ **Interior – Walls**

Walls are painted in sections of approximately 3 feet by 3 feet.

Painting is begun in a corner at the ceiling. Paint down the wall until entire first strip is painted. Begin painting the next strip at the ceiling, painting in sections down the wall.

Within each section, paint a 2-inch strip [1] along ceiling line, corners, and baseboards. Then paint the remainder of the section.

Paint a 2-inch strip around windows, doors, and other obstacles before painting within a section.

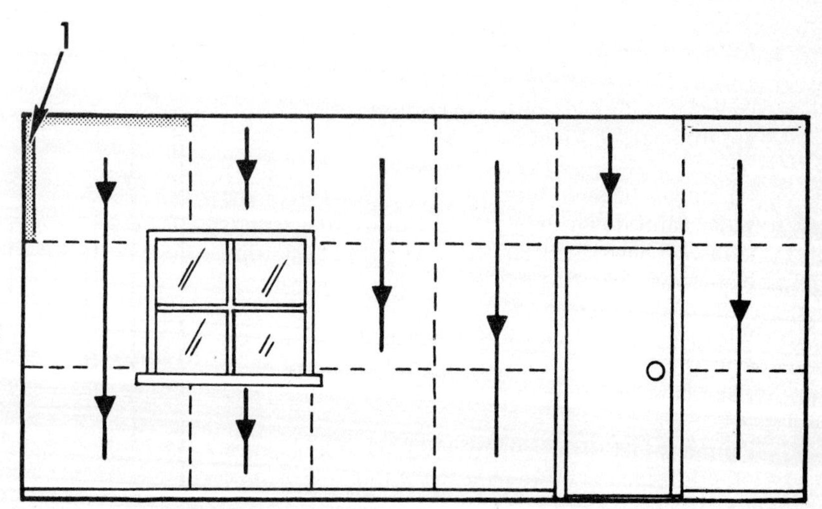

SEQUENCE OF PAINTING

▶ Exterior Sequence

Instructions in this section describe the sequence to be followed for exterior painting.

Exterior painting is done in the following sequence, with eaves painted first and accessories painted last:

- Eaves and porch ceilings, Page 84.
- Walls, Page 84.
- Trim and porch rails
- Gutters and downspouts
- Windows, Page 85.
- Doors, Page 87.
- Accessories

▶ Exterior – Eaves and Porch Ceilings

When painting eaves and porch ceilings, do not apply a thick coat of paint. These surfaces are in sheltered areas and paint is not exposed to wear from weather. If layers of paint are thick, and are applied as frequently as paint on rest of house, paint will eventually develop cracks.

▶ Exterior – Walls

Exterior walls are painted in sections of approximately 3 feet by 3 feet.

Painting is begun in a corner [1] at the top of the wall. Paint down the wall until entire first strip is painted. Begin painting the next strip at the top of the wall, painting in sections down the wall.

Within each section, paint a 2-inch strip at top of wall at corners if required, and at bottom of wall. Then paint the remainder of the section.

Paint a 2-inch strip around windows, doors, and other obstacles before painting within a section.

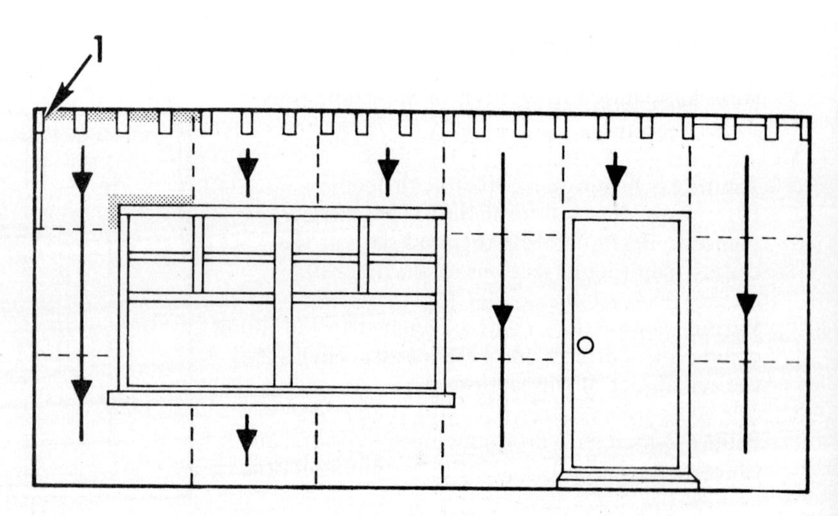

▶ Exterior — Frame Windows

Horizontal muntins are particularly subject to damage from moisture. They should be sanded before painting.

1. Raise inside sash [1] to approximately 3/4-open position. Lower outside sash [7] to approximately 3/4-open position.

When painting window, be sure to apply paint 1/16-inch onto glass to make an airtight seal.

2. Paint outside sash as follows:

   Horizontal muntins [5]
   Vertical muntins [6]
   Top sash and top edge [7]
   Two vertical sashes [8]
   Bottom sash [9]

3. Paint inside sash as follows:

   Horizontal muntins [1]
   Vertical muntins [2]
   Top sash [3]
   Two vertical sashes [4]

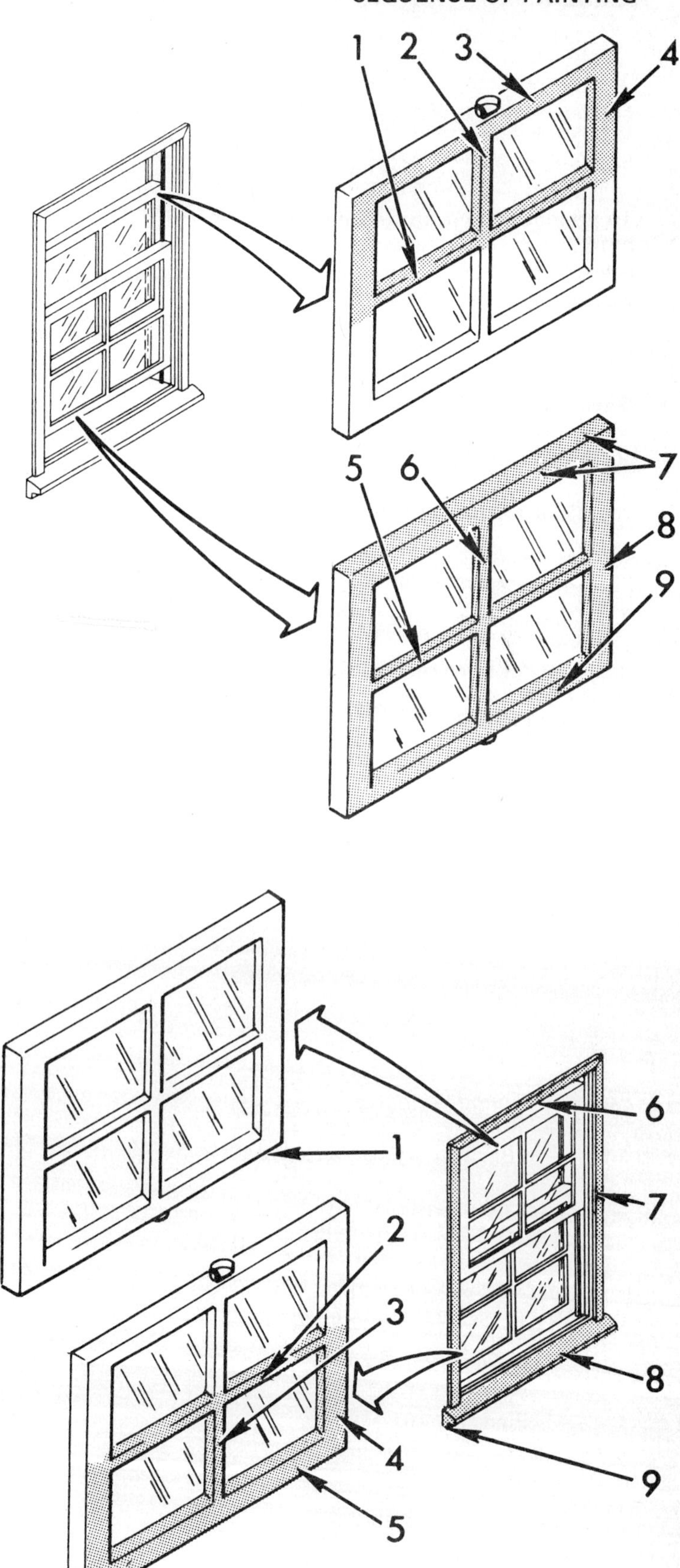

Exterior – Frame Windows

4. Lower inside sash to approximately 1 inch from fully closed position. Raise outside sash to approximately 1 inch from fully closed position.

When painting window, be sure to apply paint 1/16-inch onto glass to make an airtight seal.

5. Paint remaining surfaces of inside sash as follows:

   Horizontal muntins [2]
   Vertical muntins [3]
   Two vertical sashes [4]
   Bottom sash [5]

6. Paint bottom edge [1] of outside sash.

7. Paint frame as follows:

   Head casing [6]
   Two side casings [7]
   Sill [8]
   Apron [9]

SEQUENCE OF PAINTING

Exterior – Frame Windows

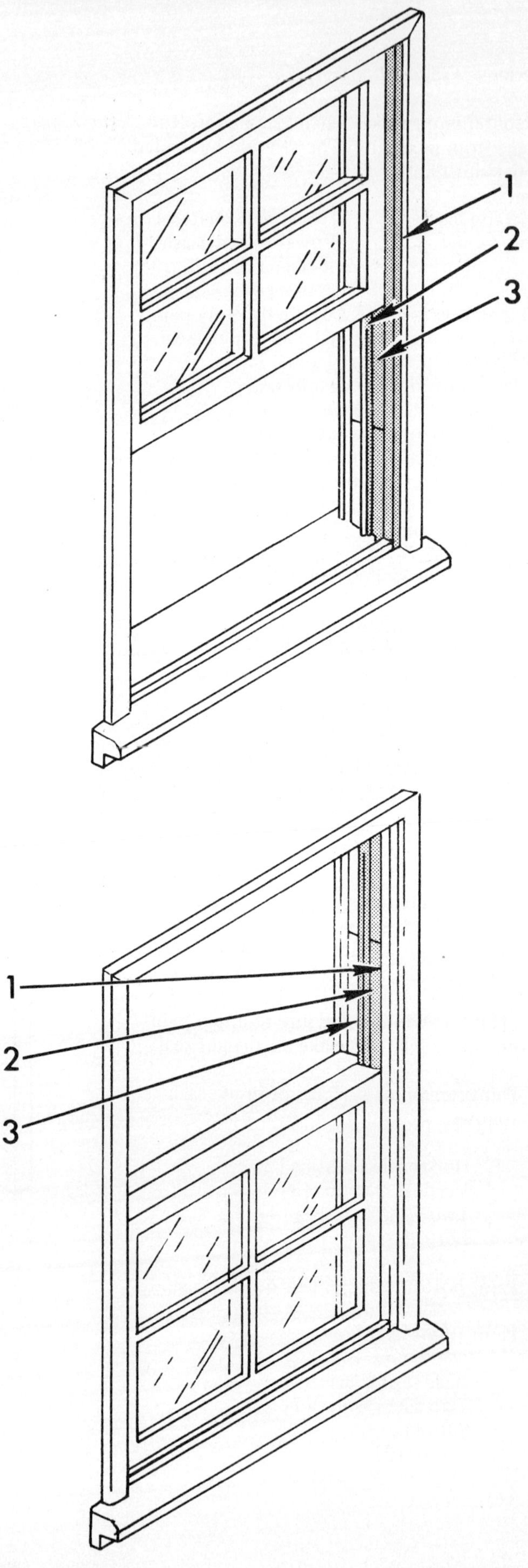

CAUTION

To prevent them from sticking, inside sash and outside sash should be moved frequently while paint is drying. When moving sashes, do not close them completely.

8. Allow paint to dry completely.

When painting window jambs, only the surfaces which are seen when the window is fully closed are to be painted.

9. Paint top half of jamb as follows:

   Two visible sides of rear stop [1]

10. Paint bottom half of jamb as follows:

   All three sides of rear stop [1]
   Visible side of bind stop [2]
   Channel [3]

Exterior – Frame Windows

11. Allow paint to dry completely.

12. Lower inside sash to fully closed position. Lower outside sash to fully open position.

13. Using beeswax or paraffin, rub top half of jamb of outside sash as follows:

   Inside surface of rear stop [1]
   Channel [2]
   Bind stop [3]

▶ **Exterior – Casement Windows**

1. Open window to fully open position.

When painting window, be sure to apply paint 1/16-inch onto glass.

When painting window, do not paint edge [6] of vertical strip facing the inside.

2. Paint window as follows:

   Horizontal strips [3] and edges [2]
   Vertical strips [4] and edges [5]

3. Paint frame as follows:

   Horizontals [1]
   Verticals [7]

4. Paint sill [8].

5. Paint apron [9].

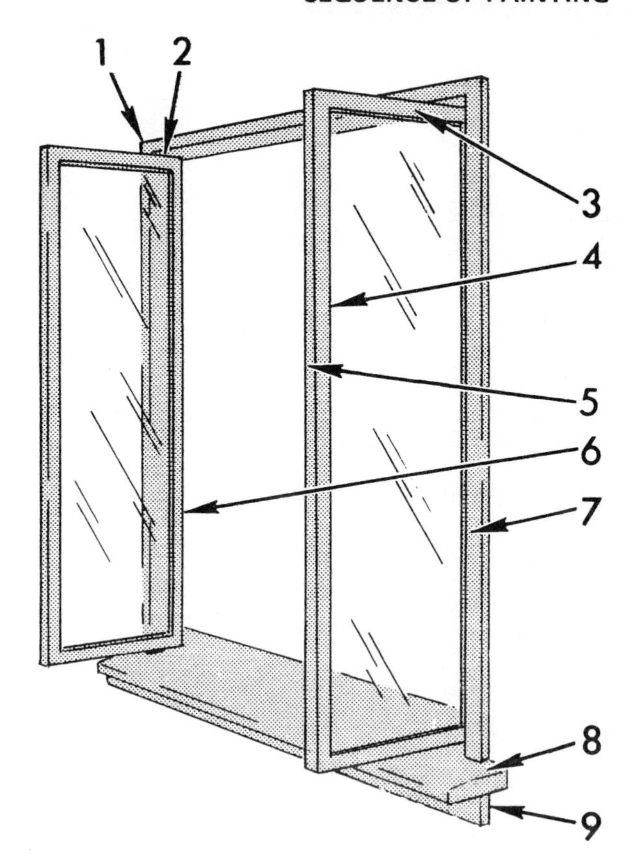

▶ **Exterior – Doors**

1. Paint door as follows:

   Inside edges [2] of panel cavities
   Panels [3]
   Horizontal strips [5]
   Vertical strips [4]
   Top edge [1]

2. If door is removed, paint bottom edge [6].

3. If door opens out, paint latch edge [8].

4. If door opens in, paint hinge edge [7].

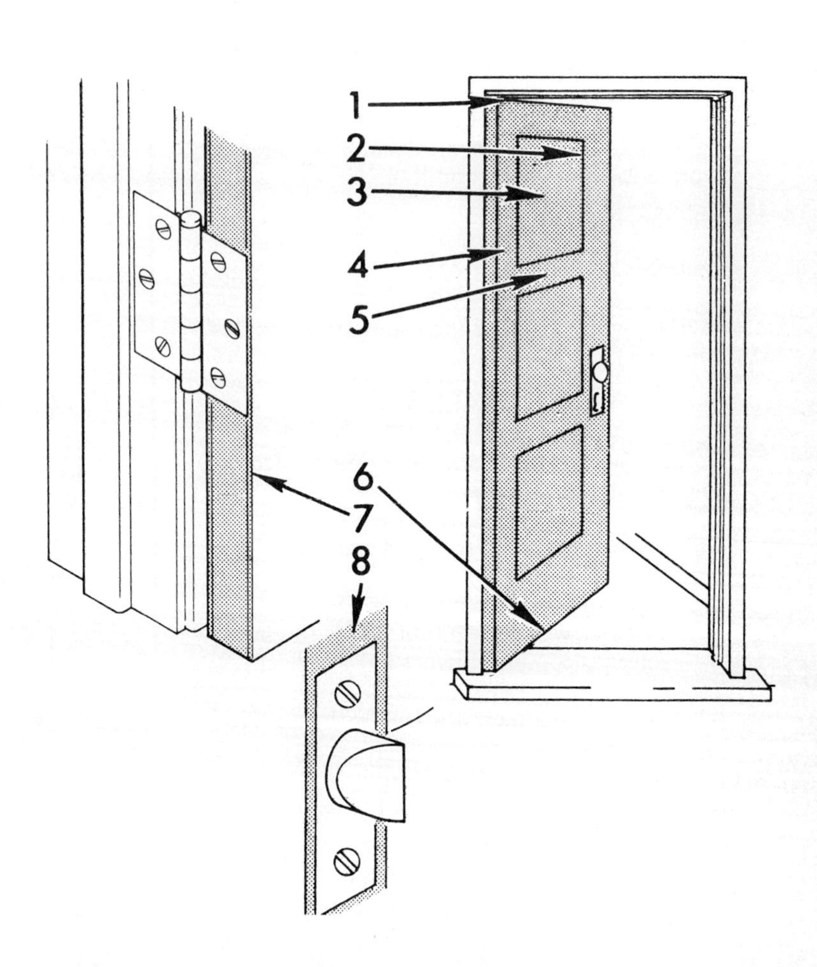

SEQUENCE OF PAINTING

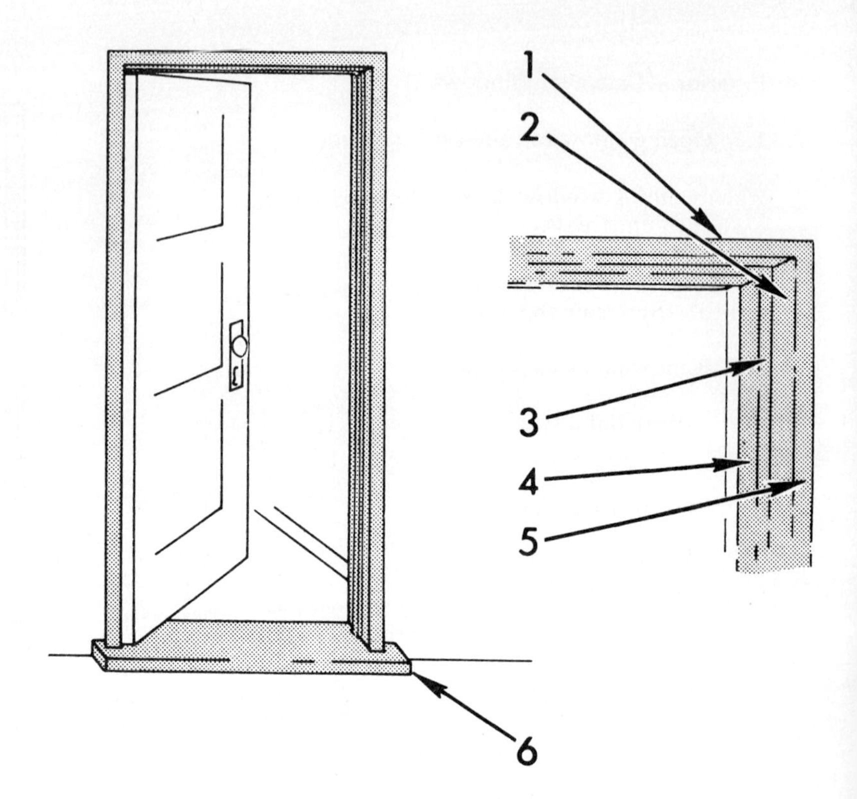

Exterior – Doors

5. Paint frame as follows:

   Top frame [1]
   Two side frames [5]

If door opens out, perform Step 3. If door opens in, perform Step 2.

6. Paint jamb [2]. Paint two sides [3,4] of door stop.

7. Paint jamb [2]. Paint door side [3] of door stop.

8. Paint threshold [6].

CLEANUP

Instructions in this section tell how to clean painting tools and work area, and how to return work area to original condition.

▶ **Cleaning Brushes**

If cleaning latex paint from brushes, go to Step 4.

If cleaning paints other than latex from brushes, continue.

WARNING

Do not use solvents near a fire or flame. Do not smoke when using solvents.

Use solvents in a well ventilated area only.

1. Pour solvent into container. Dip brush into solvent. Wash brush thoroughly forcing solvent deep into bristles [1] and heel [2].

When solvent becomes saturated with paint, it must be replaced.

2. Repeat Step 1 until solvent remains clear.

3. Remove excess solvent from bristles by shaking brush or wiping with clean rag.

If solvent is flammable, dispose of rag or store in covered metal container. Go to Step 5.

Cleaning Brushes

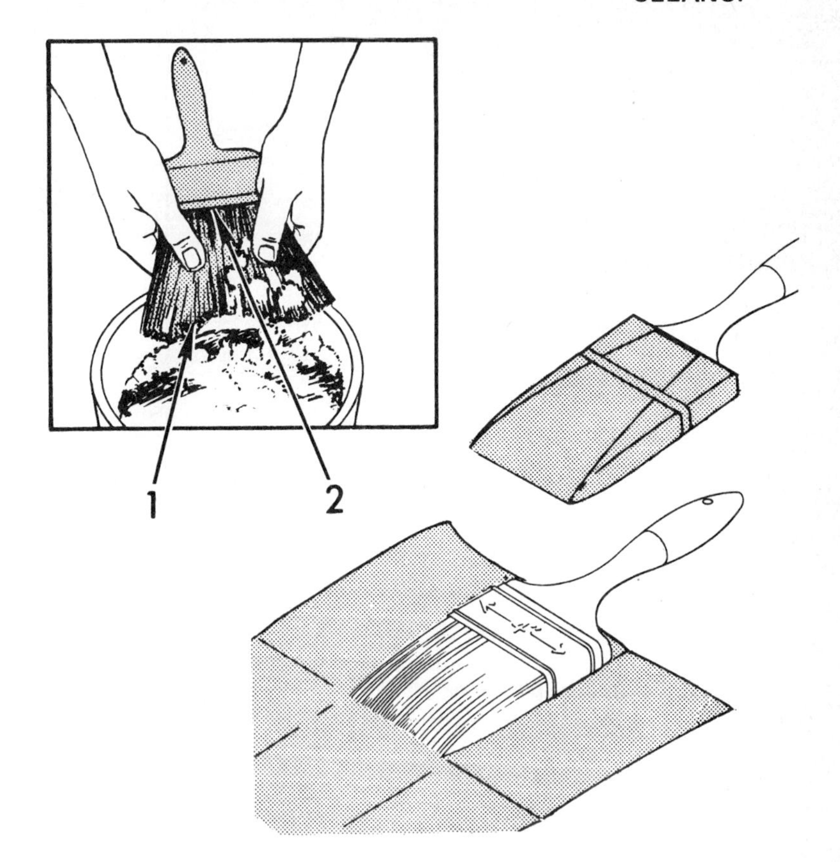

4. Hold brush under running water until water runs clear.

5. Mix a solution of soap and warm water. Wash brush in solution forcing solution into bristles [1] and heel [2].

6. Rinse brush in warm running water until water runs clear.

7. Using fine tooth comb, comb bristles thoroughly. Allow to dry completely.

8. Wrap brush in aluminum foil or plain wrapping paper.

9. Store brush by laying flat or hanging by handle.

▶ **Cleaning Rollers and Spreaders**

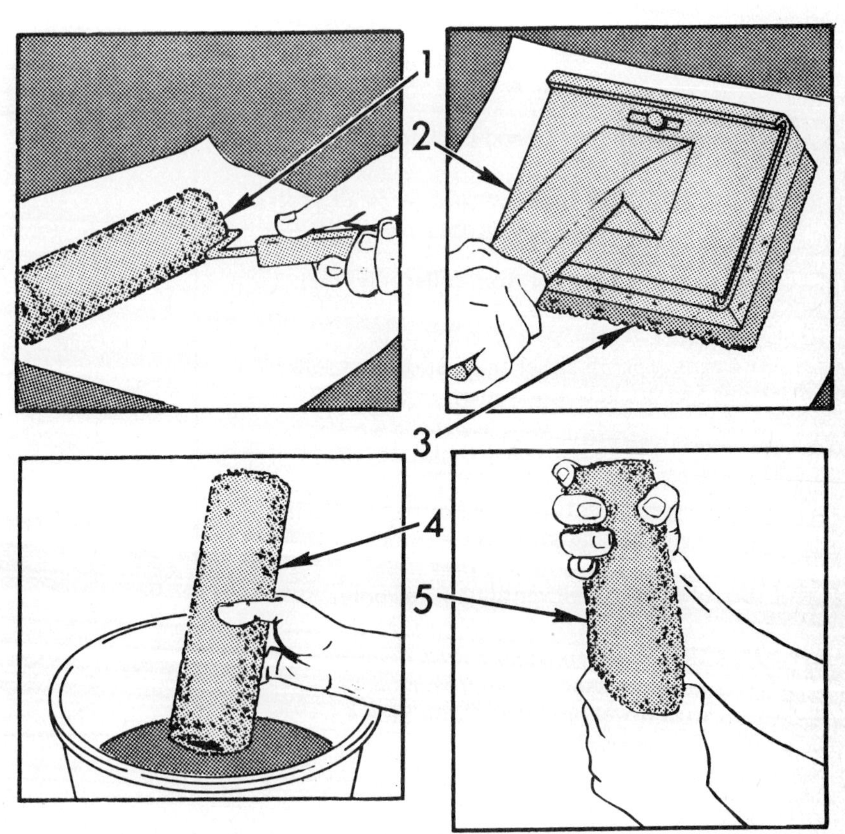

1. Remove excess paint from roller [1] or spreader [2] by pressing on a newspaper.

2. Remove roller pad [4] or spreader pad [3] from frame. Disassemble frame as required to clean parts.

If cleaning latex paint from rollers and spreaders, go to Step 6.

If cleaning paints other than latex from rollers and spreaders, continue.

WARNING

Do not use solvents near fire or flame. Do not smoke when using solvents.

Use solvents in a well ventilated area only.

3. Pour solvent into container. Dip pad into solvent. Wash pad thoroughly forcing solvent deep into nap.

When solvent becomes saturated with paint, replace solvent.

4. Repeat Step 3 until solvent remains clear.

5. Squeeze excess solvent from pad [5]. Wash frame in solvent. Go to Step 7.

Cleaning Rollers and Spreaders

6. Hold pad [1] and frame under running water until water runs clear.

7. Mix a mild solution of soap and warm water. Wash pad [2] and frame in solution forcing solution into nap.

8. Rinse pad and frame in warm running water until water runs clear.

9. Squeeze excess water from pad. Rub pad vigorously with clean absorbent cloth. Allow to dry completely. Dry frame and assemble as required.

10. Wrap pad in aluminum foil, plain wrapping paper [3] or plastic bag.

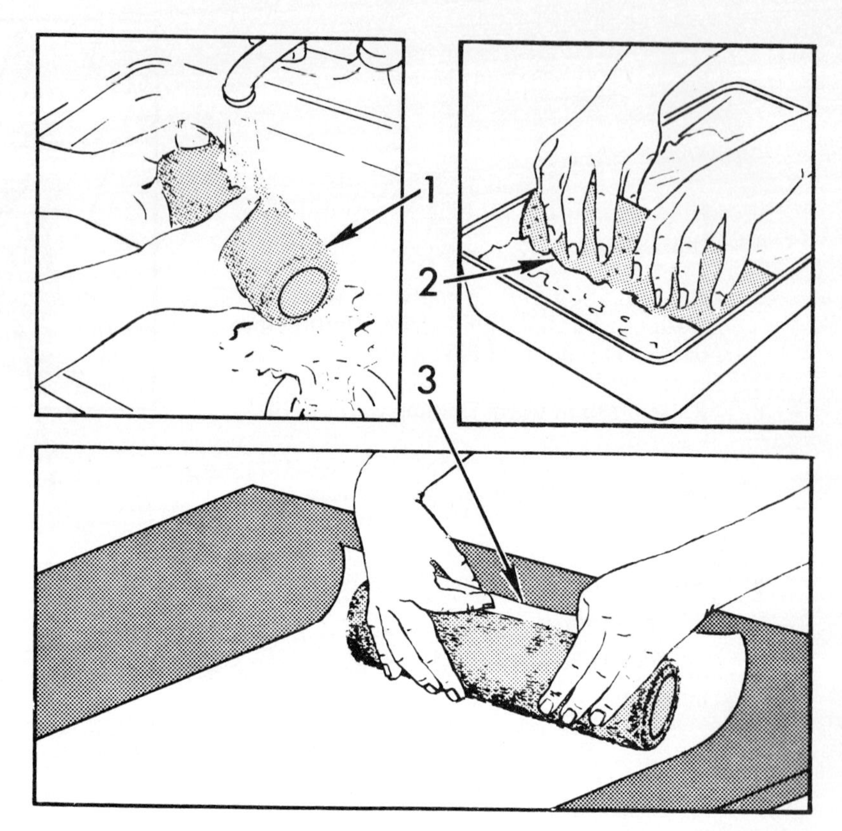

▶ **Cleaning Roller Trays and Buckets**

1. Using paper towels or rags, wipe all excess paint from bucket or tray.

If cleaning latex paint from roller trays and buckets, go to Step 3.

If cleaning paints other than latex from roller trays and buckets, continue.

WARNING

Do not use solvents near fire or flame. Do not smoke when using solvents.

Use solvents in well ventilated area only.

2. Moisten clean paper towels or cloth with solvent. Wipe all paint from bucket or tray with towel or cloth. Go to Step 4.

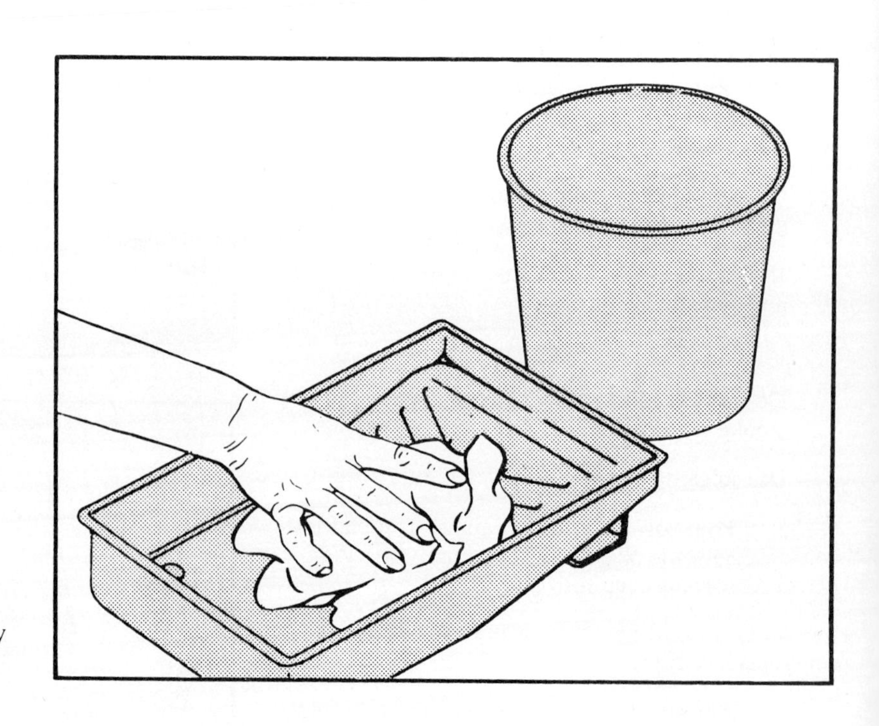

Cleaning Roller Trays and Buckets

3. Hold bucket or tray under warm running water until water runs clear.

4. Mix a mild solution of soap and water. Wash bucket or tray in solution.

5. Rinse well in warm running water until water runs clear.

6. Dry bucket or tray with clean, absorbent cloth or paper towels.

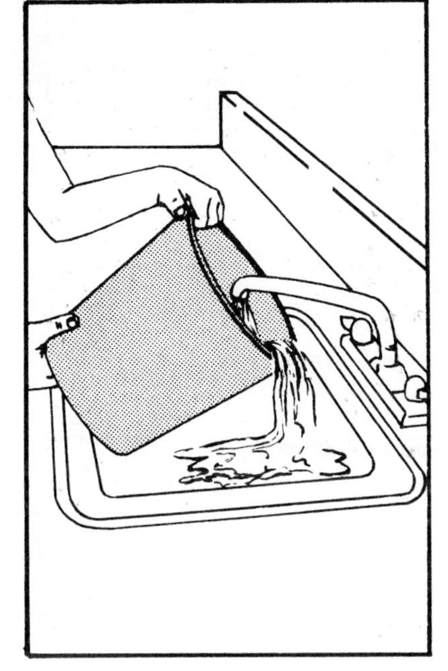

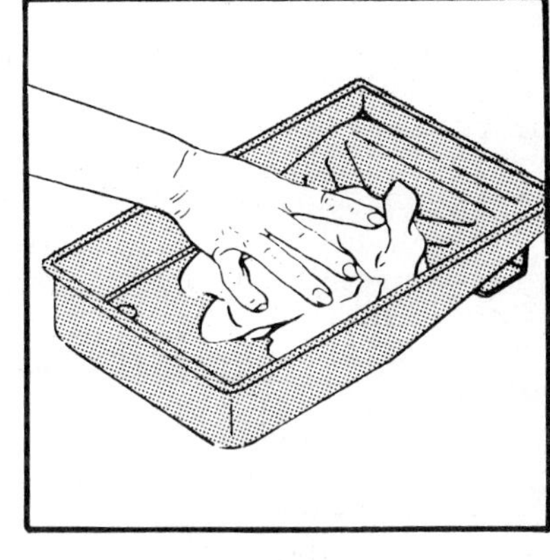

▶ Interior Cleanup

1. Remove all masking tape and plastic bags from windowpanes and fixtures.

2. Install all fixtures on doors, windows, and cabinets.

3. Install doors if removed.

4. Install wall fixtures such as curtain rods and switch covers. Install curtains and pictures.

5. Wipe excess paint from drop cloths. Allow to dry completely. Fold and store drop cloths.

CLEANUP

▶ Exterior Cleanup

1. Remove all masking tape and plastic bags from windowpanes and accessories.
2. Install all fixtures on doors and windows.
3. Install doors if removed.
4. Install house numbers, mailbox, shutters, screens, and other accessories.
5. Wipe excess paint from drop cloths. Allow to dry completely. Fold and store drop cloths.
6. Untie shrubs and bushes.

PAINT PROBLEMS AND REPAIRS

Instructions in this section show how to identify and repair different kinds of paint problems around the home.

Be sure to read Page 93 (top) before beginning any repairs. It describes how to feather a painted surface into a bare area, an important step in making an area smooth for painting.

▶ Tools and Supplies

In some cases, surfaces can be prepared by removing loose, flaking paint. In other cases, surface preparation requires that paint be removed down to bare wood.

When removing old paint, be careful not to scratch or gouge the surface.

The following tools and supplies may be used:

Scraper [1]
Wire brush [2]
Drill motor with wire brush [3] or
  sanding disc [4]
Power sander [5]
Sanding block [6]

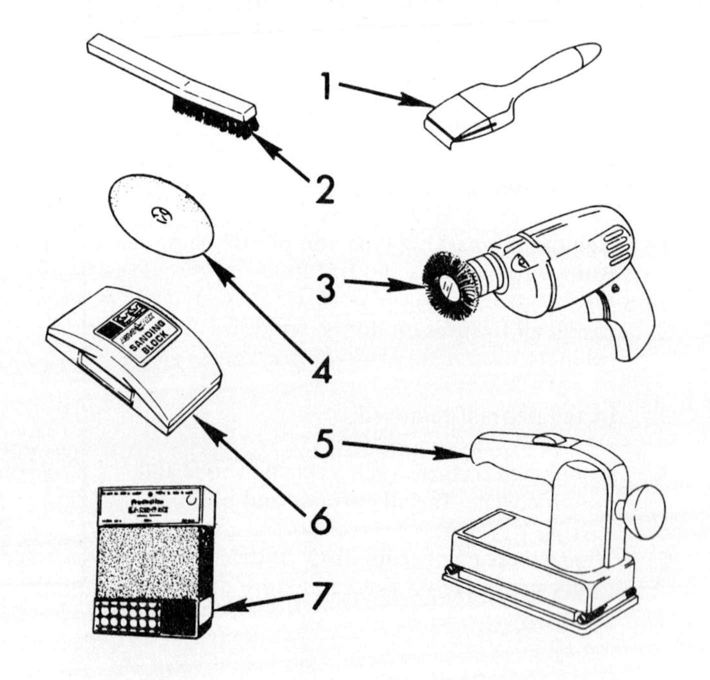

Use good quality coarse grit sandpaper [7]. Sandpaper comes in several grades and minerals. Aluminum oxide sandpaper gives longer life and faster cutting.

▶ Feathering

Feathering is a sanding technique which is done after the paint is removed down to the bare surface.

Feathering makes the surrounding good layer of paint [1] slope gradually into the bare area [2].

Feathering will make the paint stick better and the repair will not show as much.

Do not use power tools or sanding block when feathering. Sand only by hand.

Quality sandpaper should be used.

Using medium grit sandpaper, gently sand edges of good paint [1] until they gradually slope into bare area [2].

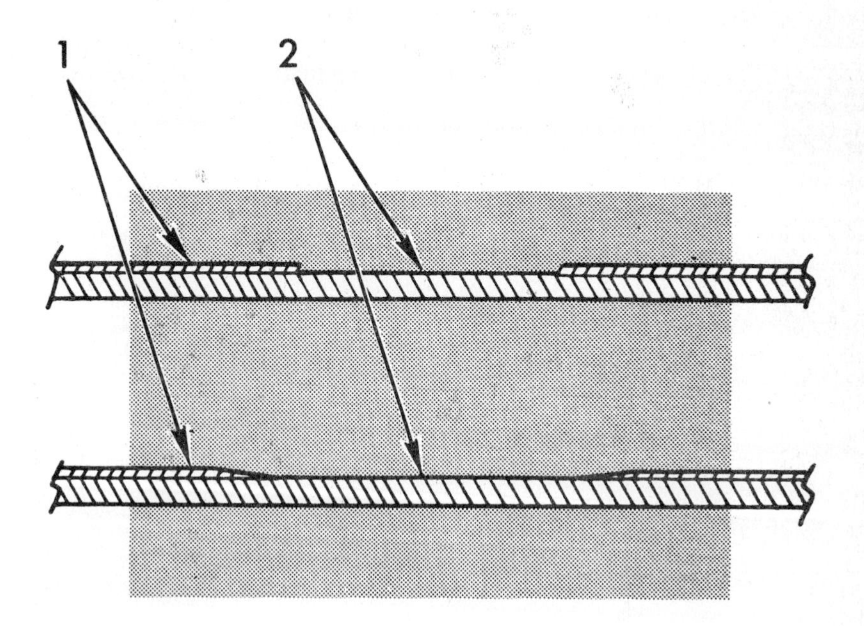

▶ Repairing Cracking and Alligatoring

Cracking and alligatoring [1] are caused by:

- Paint coat too thick
- Too many coats of paint
- Paint applied over paint coat which is not completely dry

The following tools and supplies are required:

Tool to remove loose, flaking paint, Page 92.
Tool to remove paint to bare surface, Page 92.
Medium grit sandpaper
Paintbrush
Undercoat or primer

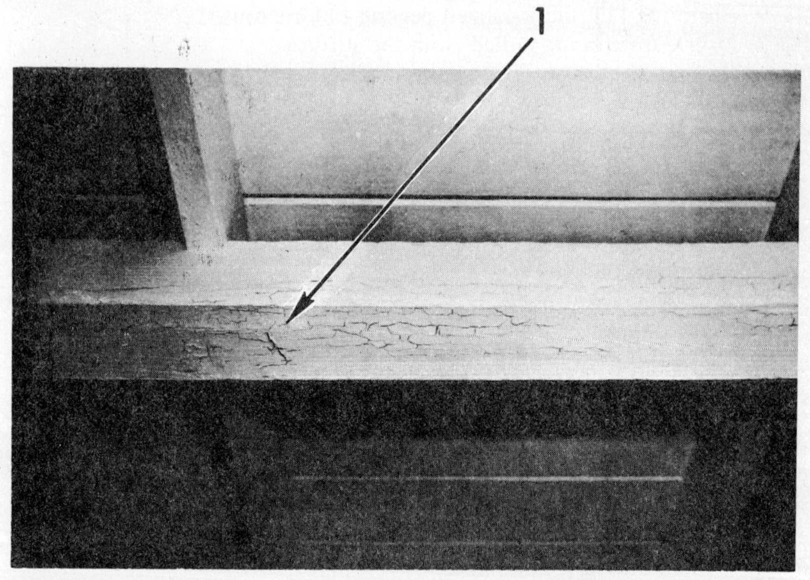

1. Remove all loose, flaking paint from damaged area.
2. Sand all paint from damaged area until bare surface can be seen.
3. Feather edges of painted surface into bare area. See above for information on feathering.
4. Apply undercoat or primer to bare area. Allow to dry completely.

PAINT PROBLEMS AND REPAIRS

▶ Repairing Checking

Checking [1] is caused by expansion and contraction of a wood surface as it ages.

The following tools and supplies are required:

Tool to remove paint to bare surface, Page 92.
Medium grit sandpaper
Paintbrush
Undercoat or primer
Wood putty
Putty knife

1. Sand all paint from damaged area until bare wood can be seen.
2. Feather edges of painted surface into bare area. Go to Page 93 for information on feathering.
3. Apply undercoat or primer to bare area. Allow to dry completely.
4. Using wood putty, fill all checking cracks.
5. Apply undercoat or primer to repaired area. Allow to dry completely.

▶ Repairing Blistering, Localized Peeling and Flaking

Blistering [1] and localized peeling [1] are caused by moisture being pulled from the surface through the paint, lifting paint away from the surface.

Flaking [2] is caused by:

- Surface swelling as moisture is absorbed, and shrinking as moisture dries. The paint cracks from swelling and shrinking and flakes away from the surface.
- Paint coat too thick

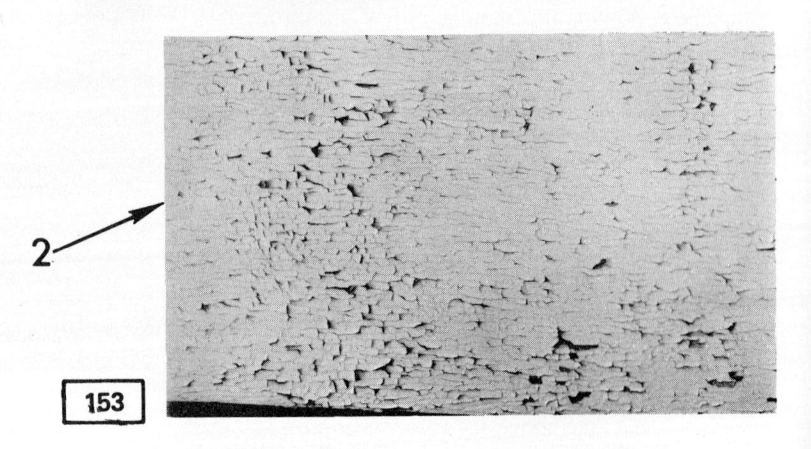

The following tools and supplies are required:

Tool to remove loose, flaking paint, Page 92.
Tool to remove paint to bare surface, Page 92.
Medium grit sandpaper
Paintbrush
Undercoat or primer

The source of moisture entering the wood must be eliminated before the surface problem is repaired.

Possible sources of moisture are plumbing and roof leaks near the damaged area. Be sure all leaks are repaired before continuing.

Repairing Blistering, Localized Peeling and Flaking

1. Remove all loose, flaking paint from damaged area [1].

2. Check that damaged area does not have a dirty gray surface. If damaged area has a dirty gray surface, note whether surface has mildew.

3. Sand all paint from damaged area and 12 inches around damaged area, until bare surface can be seen.

4. Feather edges of painted surface into bare area. Go to Page 93 for information on feathering.

If mildew was observed in Step 2, go to Page 96 to remove mildew.

If mildew was not observed, continue.

5. Apply undercoat or primer to bare area. Allow to dry completely.

▶ Repairing Topcoat Peeling

Topcoat peeling [1] is usually found on overhanging horizontal surfaces. It is caused by salt deposits which are not washed away by rain.

The following tools and supplies are required:

- Tool to remove loose, flaking paint, Page 92.
- Tool to remove paint to bare surface, Page 92.
- Medium grit sandpaper
- Paintbrush
- Undercoat or primer
- Scrub brush
- Trisodium phosphate (T.S.P.)

1. Remove all loose, flaking paint from damaged area.

2. Sand all paint from damaged area until bare surface can be seen.

3. Feather edges of painted surface and bare area. Go to Page 93 for information on feathering.

4. Mix a solution of:
   1/3-cup trisodium phosphate (T.S.P.)
   1 gallon water

5. Using scrub brush, wash bare area thoroughly with solution. Rinse well with clean water. Allow to dry completely.

6. Apply undercoat or primer to bare area. Allow to dry completely.

PAINT PROBLEMS AND REPAIRS

▶ Removing Mildew

Mildew is a fungus growth caused by:

- Moisture
- Warm temperature
- Poor ventilation

The following tools and supplies are required:

Paintbrush
Undercoat or primer
Mildew-resistant additive
Scrub brush
Trisodium phosphate
Household bleach

1. Mix a solution of:

   1/3-cup trisodium phosphate (T.S.P.)
   1/2-cup household bleach
   1 gallon warm water

2. Using scrub brush, wash mildewed area thoroughly with solution until all mildew is removed. Rinse well with clean water. Allow to dry completely.

If preparing surface for an oil base paint, be sure to add a mildew-resistant additive to undercoat or primer.

If preparing surface for a water-base paint, mildew-resistant additives are not needed.

3. Apply undercoat or primer to cleaned area. Allow to dry completely.

▶ Removing Redwood and Cedar Stains

Redwood and cedar stains are caused by water dissolving the wood coloring. The dissolved coloring flows onto the paint and, as the water dries, the stain remains on the painted surface.

The following tools and supplies are required:

Scrub brush
Denatured alcohol

The source of moisture entering the wood must be eliminated before the stains are repaired.

Possible sources of moisture are plumbing and roof leaks near the stained area. Be sure all leaks are repaired before continuing.

1. Mix a solution of:

   1 part denatured alcohol
   1 part water

2. Using scrub brush, wash stained area thoroughly with solution until all stain is removed. Rinse well with clean water. Allow to dry completely.

▶ Repairing Nailhead Stains

Nailhead stains [1] are caused by moisture rusting uncoated nailheads.

The following tools and supplies are required:

Tool to remove paint to bare surface, Page 92.
Hammer
Nail set
Paintbrush
Undercoat or primer
Wood putty
Putty knife

The source of moisture entering the wood must be eliminated before the stains are repaired.

Possible sources of moisture are plumbing and roof leaks near the stained area. Be sure all leaks are repaired before continuing.

1. Sand all paint from stained area until bare wood can be seen. Remove all stain and rust from nailhead. Be sure bare metal of nailhead can be seen.

2. Using hammer and nail set, tap nailhead 1/8-inch below surface.

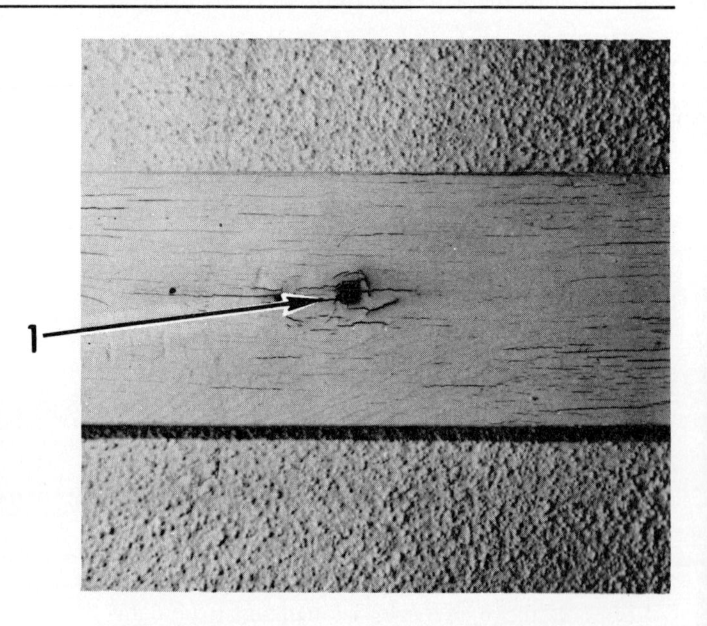

3. Apply undercoat or primer to bare area. Be sure nailhead is coated. Allow to dry completely.

4. Using wood putty, fill nailhead hole. Allow to dry Sand smooth.

5. Apply undercoat or primer to wood putty. Allow to dry completely.

▶ Repairing Peeling and Flaking Wood and Metal Gutters

Peeling and flaking wood gutters [1] are caused by the wood swelling as moisture is absorbed, and shrinking as moisture dries. The paint cracks from swelling and shrinking and flakes away from the surface.

Peeling and flaking metal gutters [1] are caused by no primer being applied to the metal, or the wrong type of primer being applied.

The following tools and supplies are required:

- Tool to remove loose, flaking paint, Page 92.
- Tool to remove paint to bare surface, Page 92.
- Medium grit sandpaper
- Roofing cement
- Heavy-gauge aluminum foil
- Paintbrush
- Undercoat or primer

If gutter is badly damaged, replacing gutters may be easier than repair.

1. Remove all loose flaking paint from damaged area.

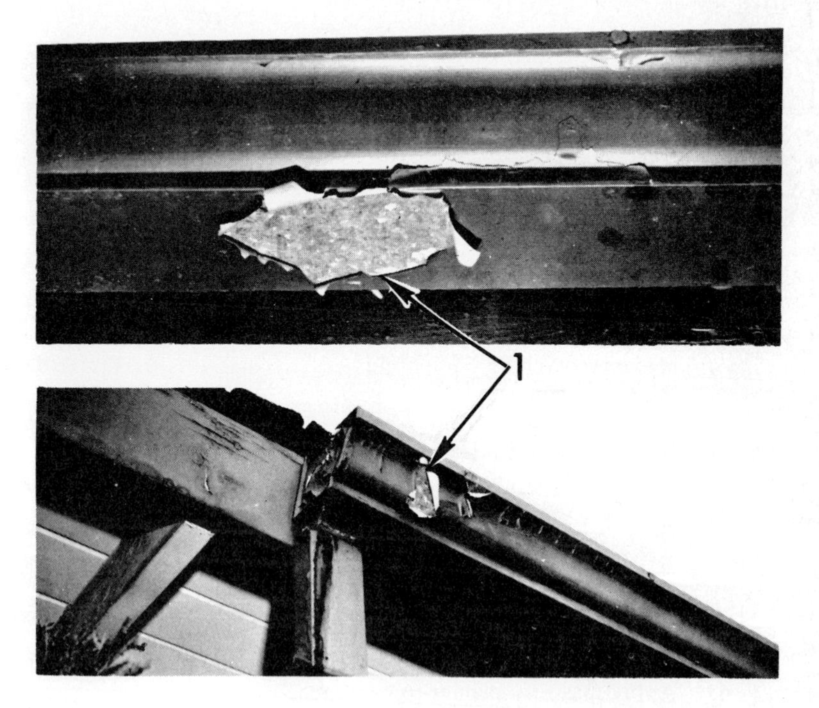

2. Remove all paint from damaged area and 12-inches around damaged area until bare surface can be seen.

Repairing Peeling and Flaking Wood and Metal Gutters

3. Feather edges of painted surface into bare area. Go to Page 93 for information on feathering.

If gutter has no holes or badly damaged areas, go to Step 8.

If gutter has holes or badly damaged areas, continue.

Roofing cement and aluminum foil patch are applied to inside surface of gutter. If pressure-sensitive foil is used, roofing cement is not required.

4. Apply roofing cement [4] to damaged area [1].

5. Cut aluminum foil patch [3] the size of damaged area [1]. Press patch tightly into cement [4] over damaged area.

6. Apply roofing cement [2] over patch [3] and first layer of cement [4].

7. Allow cement to dry according to manufacturer's instructions.

8. Apply undercoat or primer to area. Allow to dry completely.

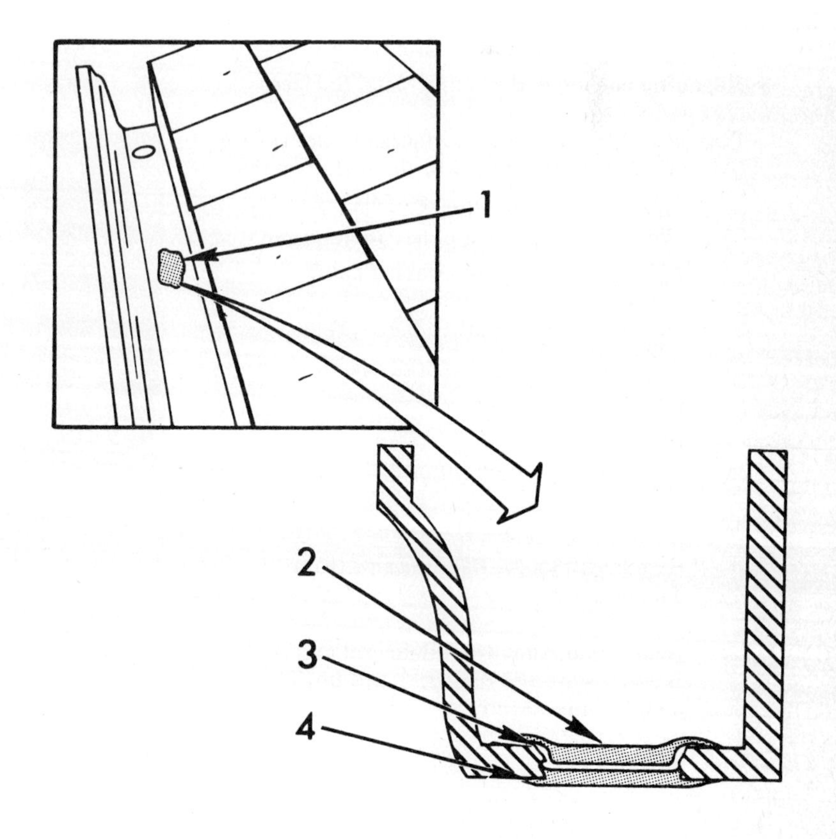

▶ Repairing Flaking and Chalking Masonry

Flaking masonry [1] and chalking masonry are caused by poor surface preparation before painting.

The following tools and supplies are required:

Tool to remove loose, flaking paint, Page 92.
Putty knife
Masonry patching compound
Masonry primer

1. Remove all loose flaking or chalking paint from damaged area.
2. Rinse bare area with clean water.
3. Using masonry patching compound, fill all cracks.
4. Apply masonry primer to bare area. Allow to dry completely.

▶ Repairing Peeling and Flaking Metal Surfaces

Peeling and flaking metal surfaces [1] are caused by no primer being applied to the metal, or the wrong type of primer being applied.

The following tools anu supplies are required:

Tool to remove loose, flaking paint, Page 92.
Tool to remove paint to bare surface, Page 92.
Medium grit sandpaper
Putty knife
Epoxy cement
Heavy-gauge aluminum foil
Paintbrush
Undercoat or primer

1. Remove all loose, flaking paint from damaged area.
2. Remove all paint from damaged area and 12 inches around damaged area until bare metal can be seen.

3. Feather edges of painted surface into bare area. Go to Page 93 for information on feathering.
4. Rinse bare area with clean water to remove all debris. Allow to dry completely.

Repairing Peeling and Flaking Metal Surfaces

If metal has no holes or badly damaged area, go to Step 9.

If metal has holes or badly damaged area, continue.

5. Apply epoxy cement [2] to damaged area [1].
6. Cut aluminum foil patch [3] the size of damaged area [1]. Press patch tightly into cement [2] over damaged area.
7. Apply epoxy cement [4] over patch [3] and first layer of cement [2].
8. Allow cement to dry according to manufacturer's instructions.
9. Apply undercoat or primer to area. Allow to dry completely.

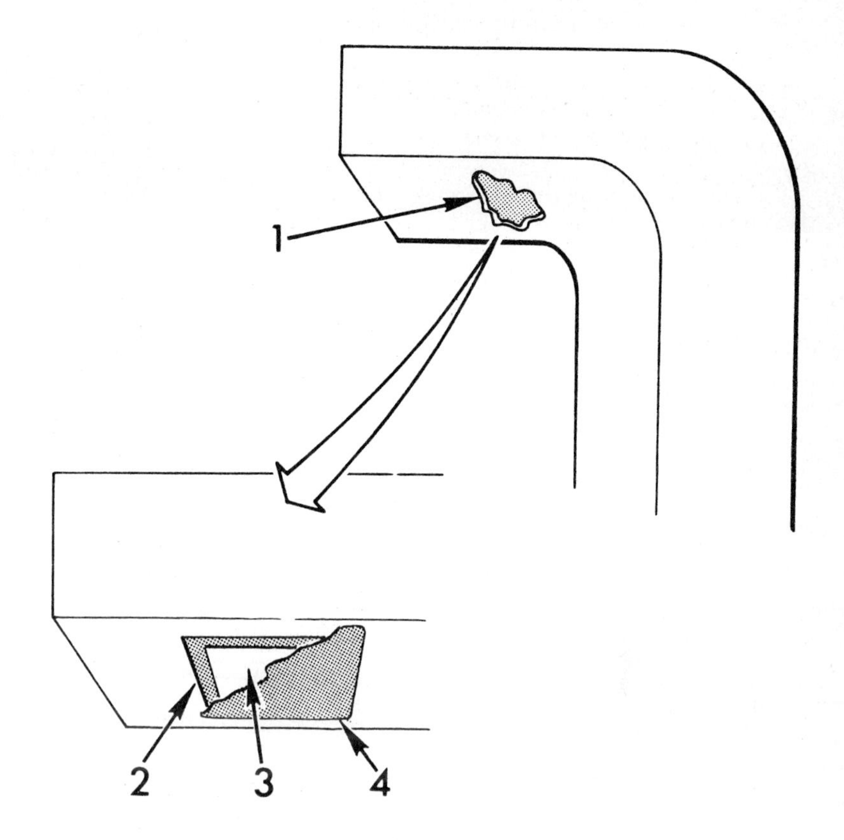

HOUSE REPAIRS

Instructions in this section show how to repair problems with stucco, plaster, and wallboard surfaces around the home. Instructions are also provided for replacing window putty.

▶ Filling Openings, Joints, or Cracks

Openings [1] are gaps between two different types of building materials such as between window frames and surrounding material.

Joints [2] are gaps between two sections of the same type of building material.

Cracks [3] are gaps in a single section of building material.

Many types of fillers are available for filling openings [1], joints [2], and cracks [3]. Ask a building supplies dealer which filler is best for a particular gap to be filled.

Caulking compound is used to fill openings [1] or joints [2]. When applying caulking compound, be sure to follow manufacturer's instructions.

Stucco patching compound is used to fill stucco cracks.

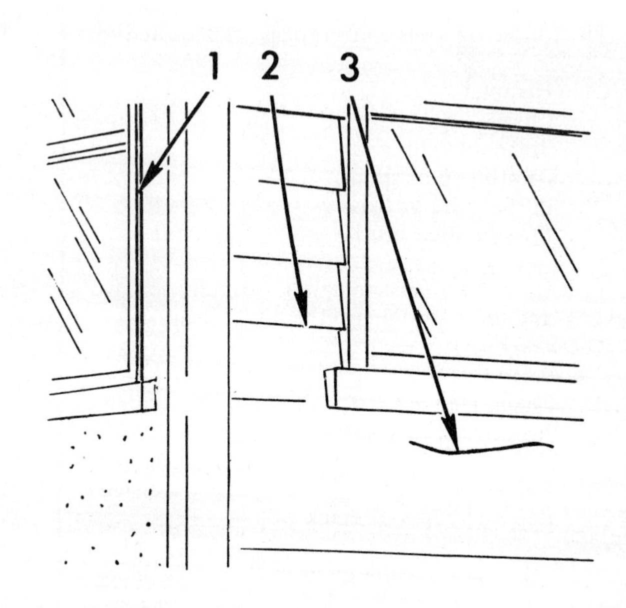

Masonry patching compound is used to fill masonry cracks.

Wood putty is used to fill wood cracks.

▶ Repairing Stucco Cracks and Openings

Cracks [3] and openings [1] in stucco surfaces [2] must be repaired to prevent structural damage to your home.

Moisture entering through cracks and openings may damage or destroy the structure. Moisture remaining in cracks and openings may freeze, making the crack or opening widen.

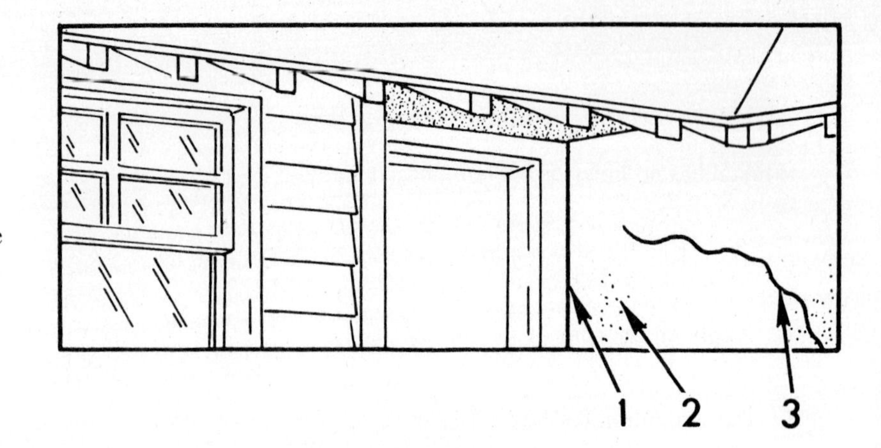

Go to next section (below) for repair of stucco cracks [3]. If repairing stucco openings [1], continue.

The following tools and supplies are required:

- Wire brush
- Medium grit sandpaper
- Paintbrush
- Undercoat or primer
- Caulking compound

When removing paint from around stucco opening, paint must be removed from both stucco and wood or metal surface.

1. Remove loose, flaking paint from stucco and wood or metal surface.
2. Sand all paint from wood or metal surface until bare wood or metal can be seen.
3. Feather edges of painted wood or metal surface into bare area. Go to Page 93 for information on feathering.
4. Apply undercoat or primer to stucco and wood or metal surface. Be sure primer goes into opening. Allow to dry completely.
5. Fill opening with caulking compound.

▶ Repairing Stucco Cracks

The following tools and supplies are required:

- Hammer
- Chisel
- Stucco patching compound
- Paintbrush or sponge
- Putty knife or trowel
- Stiff-bristled brush

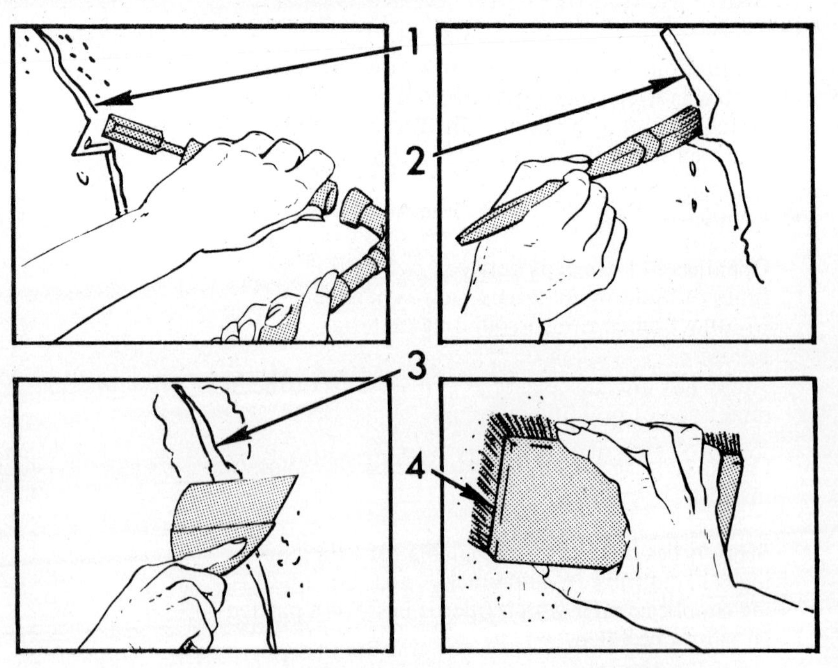

1. Using hammer and chisel, carefully remove any loose or damaged material from crack [1].
2. Mix stucco patching compound according tc manufacturer's instructions.
3. Using clean paintbrush or sponge, moisten exposed edges of crack [2].
4. Using putty knife or trowel, press patching compound into crack [3]. After crack is filled, pull putty knife or trowel along crack to make patch even with surrounding surface.
5. Using stiff-bristled brush [4], make patched crack same texture as surrounding surface.

Patching compound must dry slowly to prevent separating from surrounding stucco. To allow patching compound to dry slowly, moisten compound once every day for a week.

▶ Repairing Plaster Cracks

The following tools and supplies are required:

Paintbrush
Screwdriver or knife
Patching plaster
Sponge
Putty knife or trowel
Medium grit sandpaper
Stiff-bristled brush

1. Using dry paintbrush, remove all loose plaster from crack [1].

2. Using screwdriver, scrape out plaster beneath edges [2] of crack until crack is wider at bottom [3] than at edges.

3. Mix patching plaster according to manufacturer's instructions.

4. Using clean paintbrush or sponge, moisten plaster in and around crack.

5. Using putty knife or trowel, press plaster into crack [4].

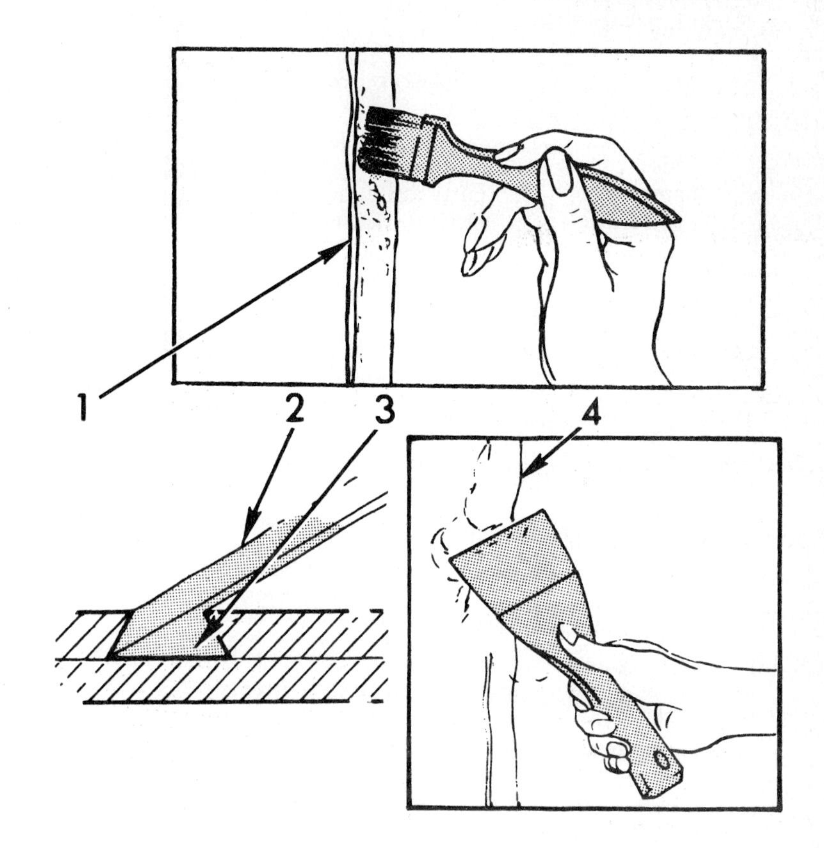

Repairing Plaster Cracks

6. Remove any excess plaster from around crack.

7. Moisten putty knife or trowel. Pull putty knife or trowel along crack [1] for smooth finish. Allow to dry completely.

If patched area is not filled, repeat Step 4 through Step 7.

If patched area is filled and is to be papered, go to Step 8.

8. Using medium grit sandpaper [2], sand plaster until level with surface.

If surface is to be painted, repair will be visible if texture of repair does not match texture of surface. Using stiff-bristled brush and thick mixture of patching compound, match texture of repair to surrounding surface.

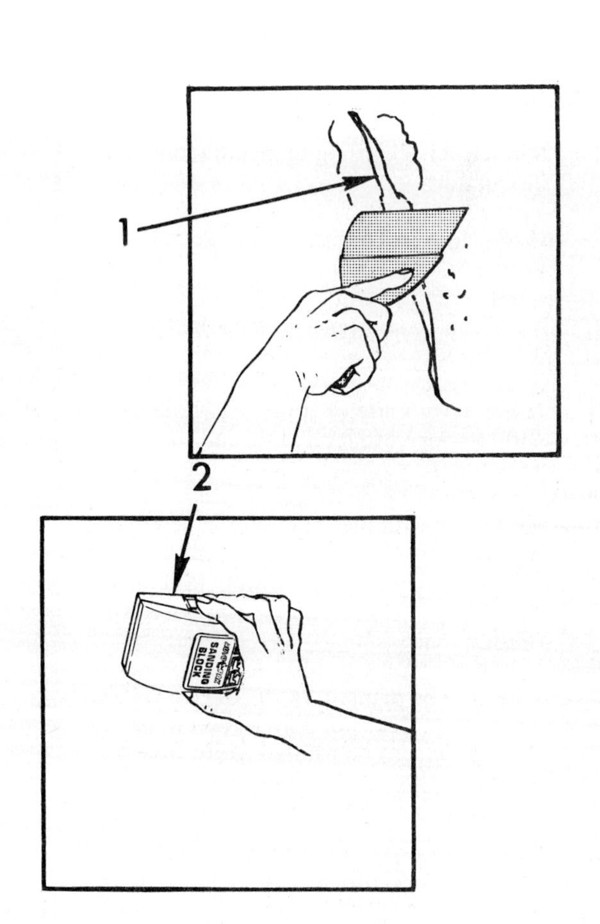

▶ Repairing Plaster Holes

The following tools and supplies are required:

Hammer	Plaster
Chisel	Sponge
Screwdriver	Putty knife or trowel
Paintbrush	Stiff-bristled brush

1. Using hammer and chisel, carefully remove any loose or damaged plaster from hole [1].

If lath [2] cannot be seen, go to Step 4.

If lath [2] can be seen, continue.

2. Check lath [2] for damage.

If lath [2] is damaged, call a professional plasterer.

If lath [2] is not damaged, continue.

3. Remove old pieces of plaster from between lath [2].

4. Using screwdriver, scrape out plaster below edges [4] of hole until hole is wider at bottom [3] than at edges.

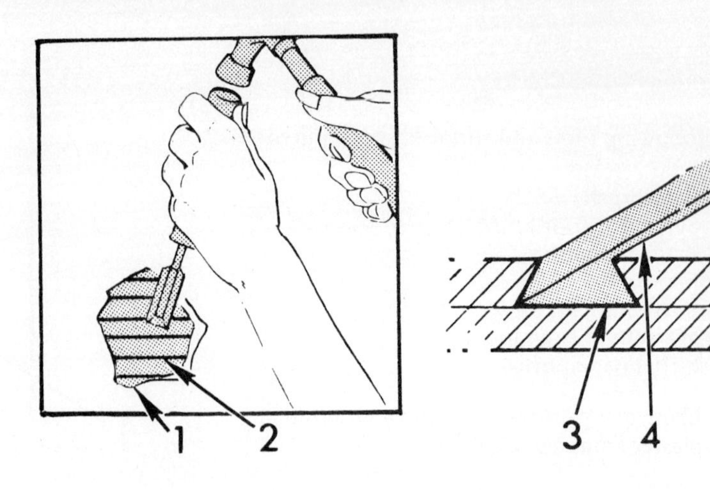

5. Using dry paintbrush, clean dust and debris from hole.

If hole is smaller than 4 inches across, you may try to repair it with one layer of plaster. Because plaster may shrink after drying, additional layers may be required. Go to next section (below).

If hole is larger than 4 inches across, it must be repaired with three layers of plaster. Go to Page 103.

▶ Repairing Plaster Holes — Holes Smaller than 4 Inches

1. Mix plaster according to manufacturer's instructions.

2. Using clean paintbrush or sponge, moisten plaster in and around hole [1].

If lath [2] can be seen, plaster must be forced around lath.

3. Using putty knife or trowel, fill hole [3] with plaster until level with wall.

Blade of putty knife or trowel should be wider than hole to smooth plaster evenly with wall.

4. Moisten putty knife or trowel. Pull putty knife or trowel across plaster for smooth finish.

If surface is to be painted, repair will be visible if texture of repair does not match texture of surface. If surface is to be painted, go to Step 5.

If surface is to be papered, go to Step 6.

5. Using stiff-bristled brush [4], match texture of repair to texture of surrounding surface.

6. Allow plaster to dry completely.

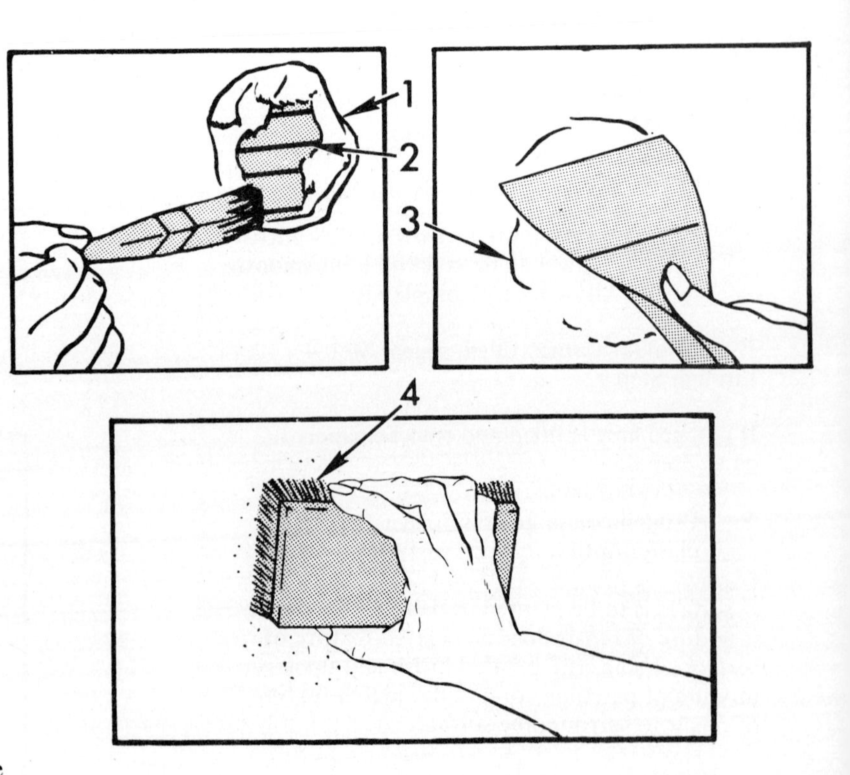

▶ Repairing Plaster Holes – Holes Larger than 4 Inches

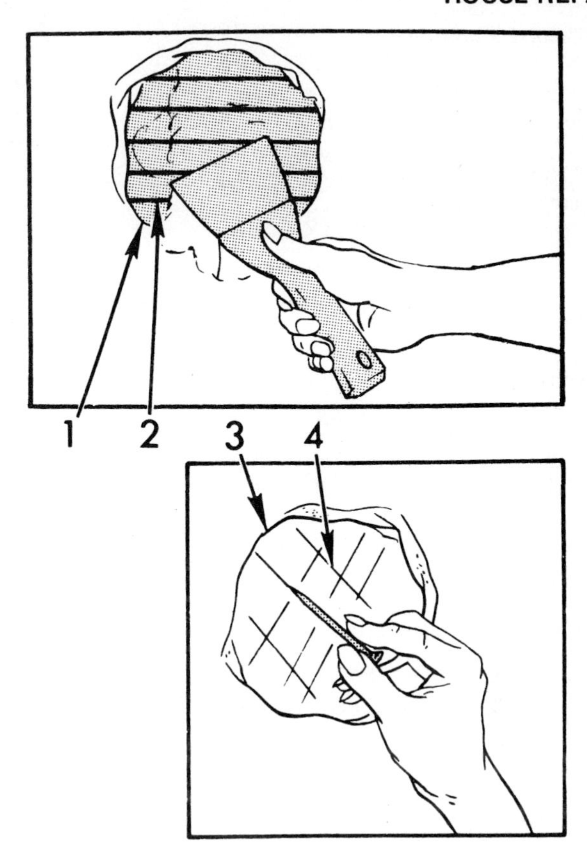

1. Mix plaster according to manufacturer's instructions.

2. Using clean paintbrush or sponge, moisten plaster in and around hole [1].

If lath [2] cannot be seen, go to Step 4.

If lath [2] can be seen, continue.

3. Using putty knife or trowel, press new plaster around lath [2].

4. Using putty knife or trowel, fill hole [1] to within 1/4-inch of surface.

5. Using nail, mark light lines [4] on new plaster. Allow to dry completely.

6. Mix plaster according to manufacturer's instructions.

7. Using clean paintbrush or sponge, moisten plaster in and around hole [3].

8. Using putty knife, fill hole [3] to within 1/8-inch of surface.

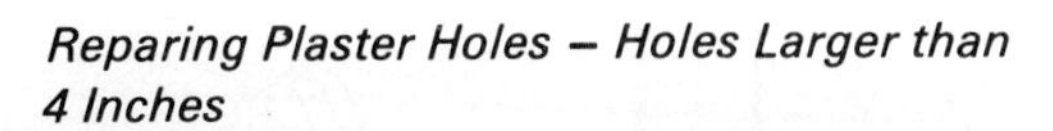

Reparing Plaster Holes – Holes Larger than 4 Inches

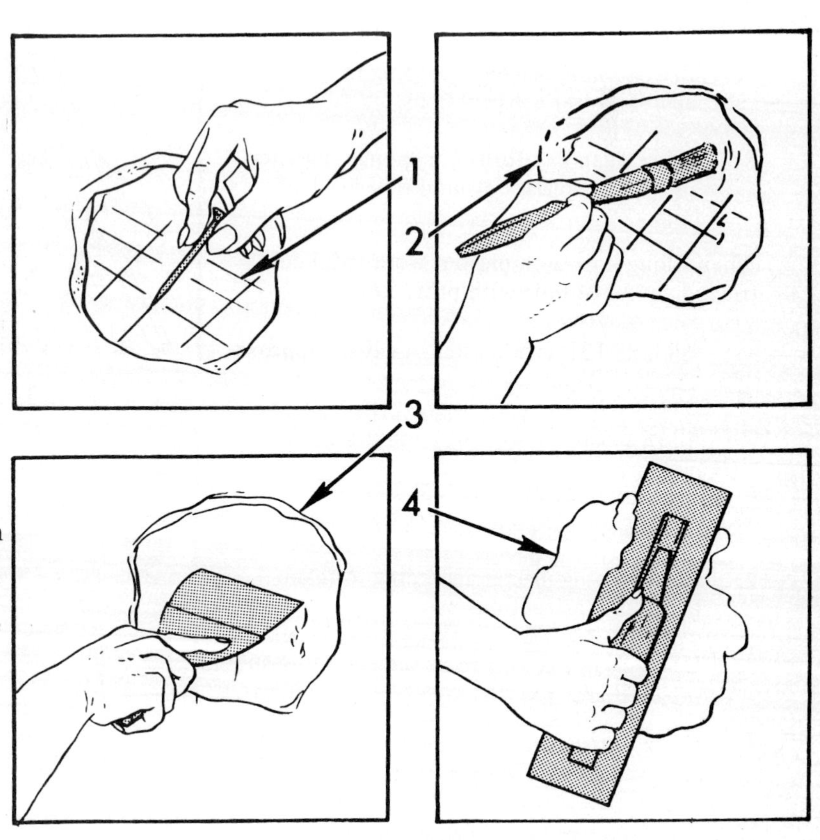

9. Using nail, mark light lines [1] on new plaster. Allow to dry completely.

10. Mix plaster according to manufacturer's instructions.

11. Using clean paintbrush or sponge, moisten plaster in and around hole [2].

Blade of putty knife or trowel should be wider than hole to aid in finishing surface.

12. Using putty knife or trowel, fill hole [3] with plaster until level with wall.

13. Moisten putty knife or trowel. Pull putty knife or trowel across plaster [4] for smooth finish.

If surface is to be painted, repair will be visible if texture of repair does not match texture of surface. If surface is to be painted, go to Step 14. If surface is to be papered, go to Step 15.

14. Using stiff-bristled brush, match texture of repair to texture of surrounding surface.

15. Allow plaster to dry completely.

▶ Repairing Wallboard Holes

These instructions apply to the repair of holes between about 1 1/2-inches and 8 inches in diameter. Smaller holes can usually be repaired by overfilling with joint cement and sanding it down after it is dry.

Holes larger than 8 inches in diameter may involve structural damage. A professional plasterer may be required.

The following tools and supplies are required:

Sharp knife	Stick, 10 inches long
Tin snips	Patching plaster
Metal lath or screen	Paintbrush or sponge
String	Putty knife or trowel

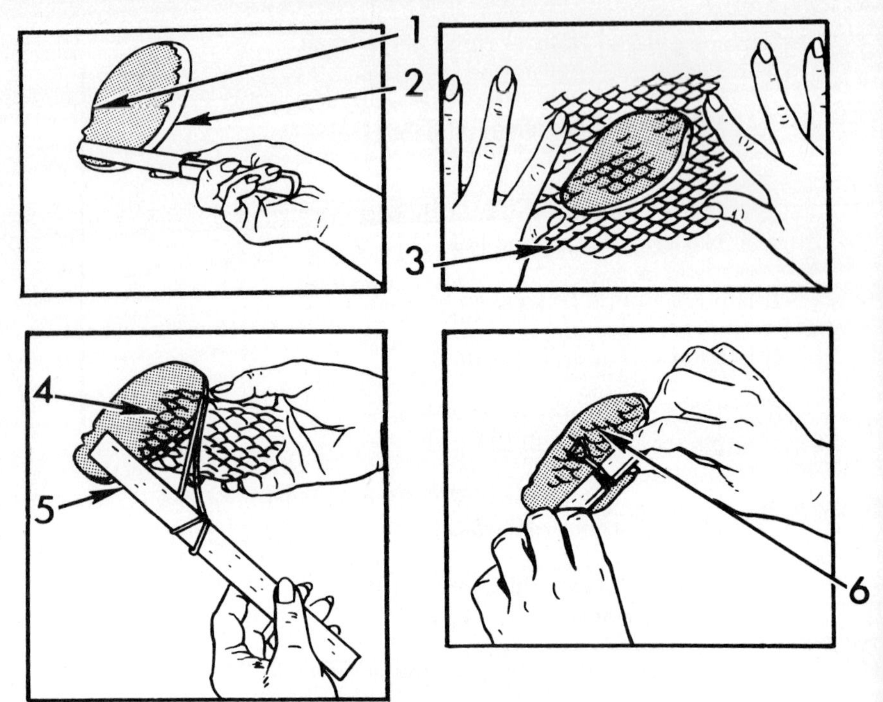

1. Cut paper from around edge of hole [2].
2. Using knife, cut all rough edges [1] of hole [2].
3. Using tin snips, cut piece of metal lath [3] or screen slightly larger than area of hole [2].
4. Push string through approximate center of lath [3] and back out again. Tie string to stick [5].
5. Roll lath [4] until small enough to insert into hole. Insert lath into hole.
6. Pull lath [6] against back of wall. Rotate stick until string is tight.

Repairing Wallboard Holes

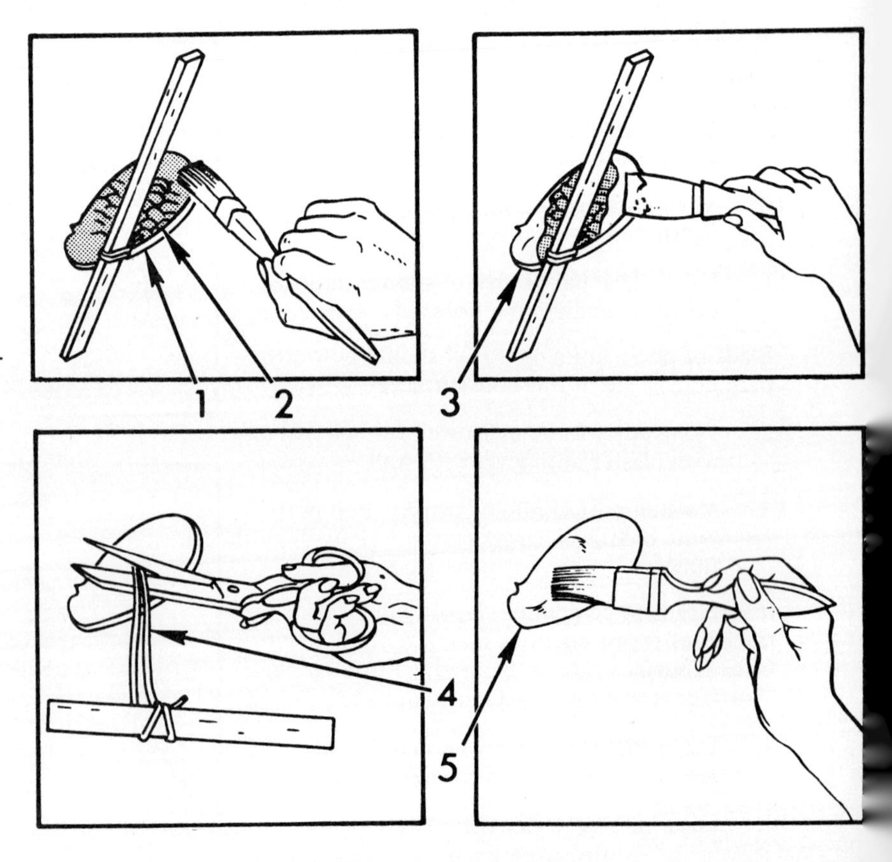

7. Mix patching plaster according to manufacturer's instructions.
8. Using clean paintbrush or sponge, moisten all exposed plaster around edges [1] of hole [2].

When filling hole with plaster, work from edges toward center of hole with putty knife.

9. Fill hole [3] with plaster to within approximately 1/4-inch of surface.
10. Allow plaster to dry for 24 hours.
11. Cut string [4] even with plaster.
12. Mix patching plaster according to manufacturer's instructions.
13. Using clean paintbrush or sponge, moisten repaired area [5].
14. Using putty knife, fill hole to within approximately 1/8-inch of surface.
15. Allow plaster to dry for 24 hours.

Repairing Wallboard Holes

16. Mix patching plaster according to manufacturer's instructions.

17. Using clean paintbrush or sponge, moisten repaired area [1].

18. Using putty knife or trowel, fill hole [2] with plaster until level with wall.

Blade of putty knife or trowel should be wider than hole to smooth plaster evenly with wall.

19. Moisten putty knife or trowel. Pull putty knife or trowel across plaster for smooth finish.

If surface is to be painted, repair will be visible if texture of repair does not match texture of surface. If surface is to be painted, go to Step 20.

If surface is to be papered, go to Step 21.

20. Using stiff-bristled brush [3], match texture of repair to texture of surrounding surface.

21. Allow plaster to dry for 24 hours.

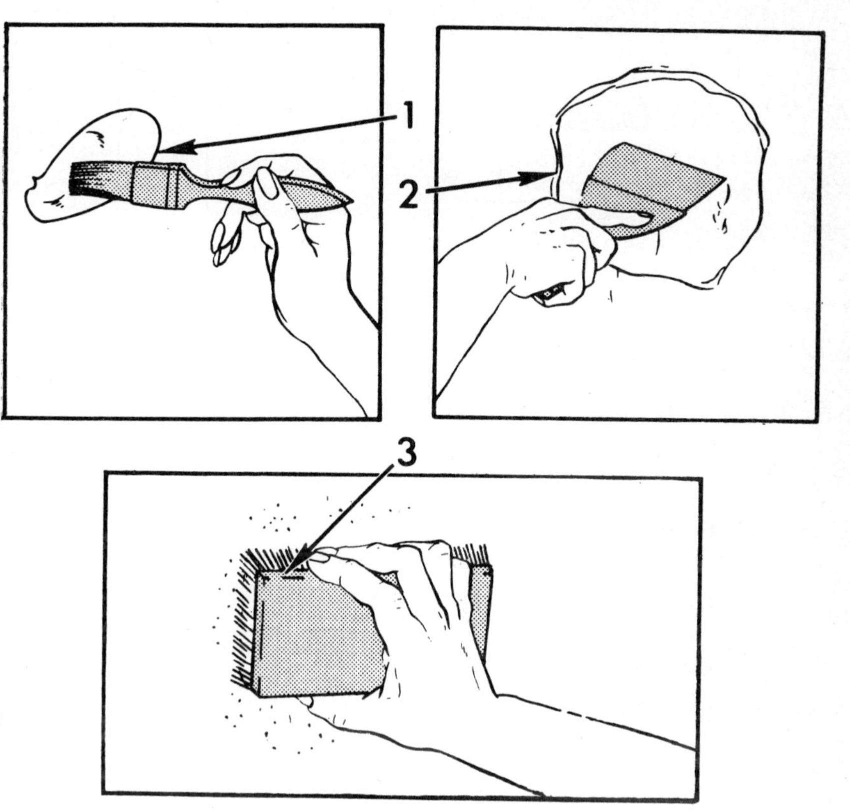

▶ Replacing Window Putty

Window putty dries out as it ages. The putty becomes cracked and separates from the surface. Moisture can enter and remain in the cracks and gaps.

Old window putty should be removed and replaced with new putty before painting to ensure an airtight, waterproof seal.

CAUTION

After applying new putty, allow putty to dry for 7 days before painting window.

The following tools and supplies are required:

- Hammer
- Chisel
- Wood rasp
- Paintbrush
- Screwdriver
- Metal primer, linseed oil, or wood preservative
- Putty or glazing compound
- Putty knife

Windowpanes [1] are held in their frames by glazier's points [3] in wooden frames, or spring clips [2] in metal frames.

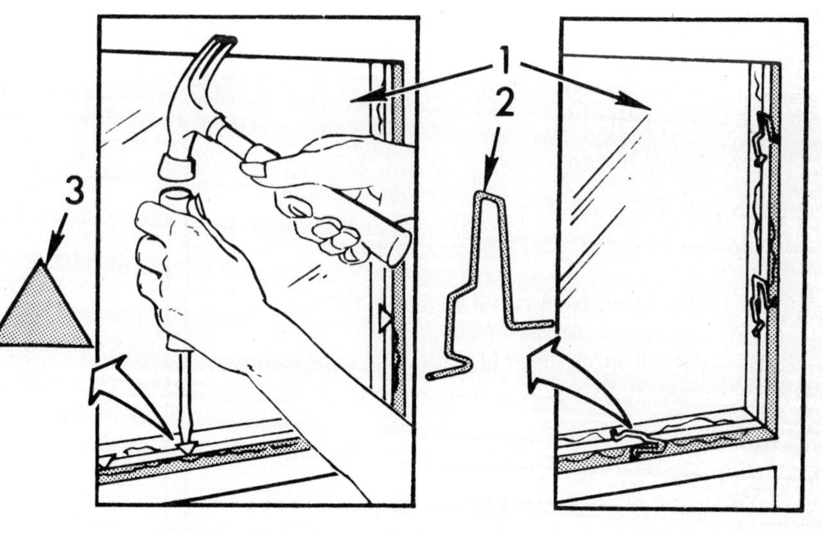

When removing old putty, the glazier's points [3] or spring clips [2] are removed as putty is removed around them.

After old putty is removed from one of the four sides of the window, the glazier's points [3] or spring clips [2] are installed to hold the windowpane [1] in place while putty is removed from the other sides.

Replacing Window Putty

CAUTION

When removing old putty, be careful not to break glass.

A warm soldering iron may be used to soften old putty, making it easier to remove.

1. Using hammer and chisel, carefully remove old putty from frame [2].
2. Remove each spring clip [3] or glazier's point [1] as old putty is removed.
3. Using wood rasp, clean all traces of old putty from frame [4].
4. Using dry paintbrush, remove all dust and debris from frame [5].

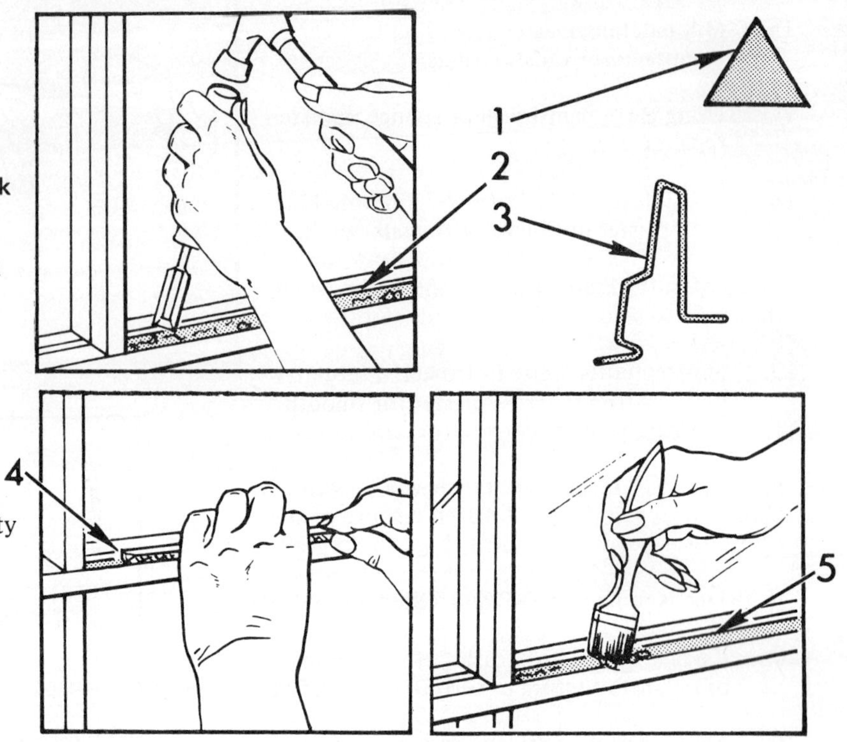

Replacing Window Putty

If replacing putty in wooden frames, go to Step 6. Be sure to read CAUTION before Step 6.

If replacing putty in metal frames, continue.

5. Install spring clips [1] allowing clips to press against glass [2]. Go to Step 7.

CAUTION

When installing glazier's points, be careful not to break glass.

6. Using hammer and screwdriver, install glazier's points [3] into frame [4].
7. Repeat Step 1 through Step 6 for remaining sides.

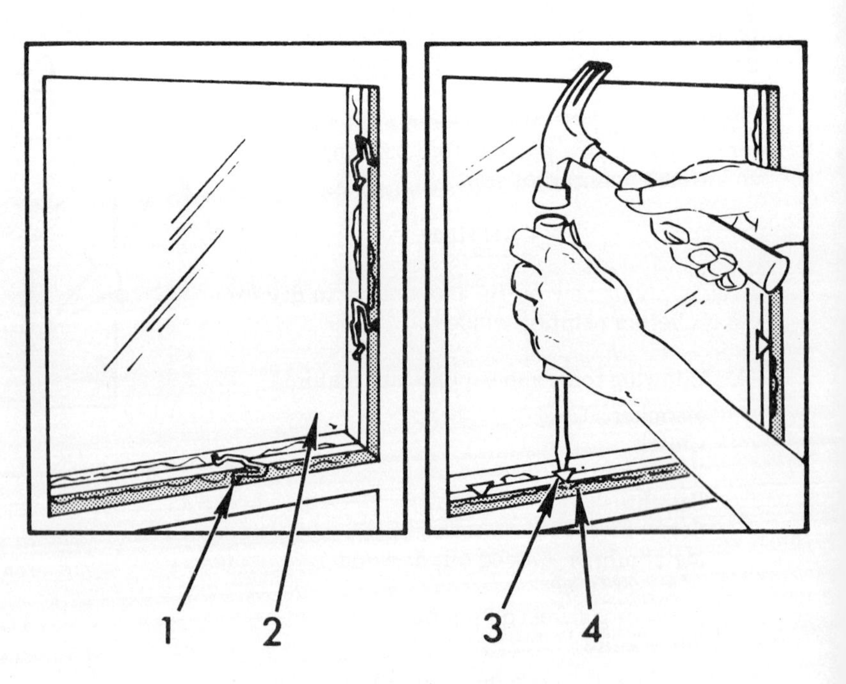

Replacing Window Putty

If replacing putty in wooden frames, go to Step 9.

If replacing putty in metal frames, continue.

8. Apply metal primer to exposed metal surfaces of frame [1]. Allow to dry completely. Go to Step 10.

9. Apply light coat of linseed oil or wood preservative to exposed wood surfaces of frame [1]. Allow to dry completely.

10. Roll putty [2] into long strings, approximately 1/4-inch thick.

11. Working rapidly, press putty [2] into frame.

Putty [3] should be even with outer edge of molding [4] and even with inner edge [5] (behind glass).

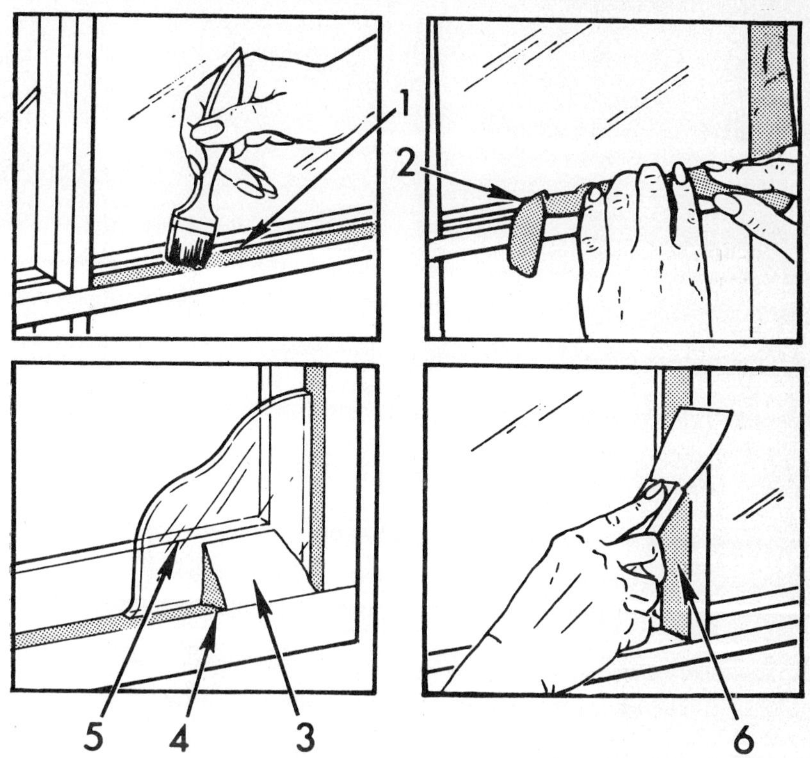

12. Pull putty knife along putty [6] until smooth finish is obtained. If putty sticks to knife, wet knife blade with water.

13. Using putty knife, trim excess putty from frame and pane.

14. Repeat Steps 10 through 13 for three remaining sides.